THE MAILLARD REACTION

RSC Food Analysis Monographs

Series Editor: P.S. Belton, *School of Chemical Sciences, University of East Anglia, Norwich, UK*

The aim of this series is to provide guidance and advice to the practising food analyst. It is intended to be a series of day-to-day guides for the laboratory worker, rather than library books for occasional reference. The series will form a comprehensive set of monographs providing the current state of the art on food analysis.

Dietary Fibre Analysis

by David A.T. Southgate, *Formerly of the AFRC Institute of Food Research, Norwich, UK*

Quality in the Food Analysis Laboratory

by Roger Wood, *Joint Food Safety and Standards Group, MAFF, Norwich, UK*, Anders Nilsson, *National Food Administration, Uppsala, Sweden*, and Harriet Wallin, *VTT Biotechnology and Food Research, Espoo, Finland*

Chromatography and Capillary Electrophoresis in Food Analysis

by Hilmer Sørensen, Susanne Sørensen and Charlotte Bjergegaard, *Royal Veterinary and Agricultural University, Frederiksberg, Denmark*, and Søren Michaelsen, *Novo Nordisk A/S, Denmark*

Mass Spectrometry of Natural Substances in Foods

by Fred A. Mellon, *The Institute of Food Research, Norwich, UK*, Ron Self, *University of East Anglia, Norwich, UK*, and James R. Startin, *Central Science Laboratory, York, UK*

The Maillard Reaction

by Siân E. Fayle, *Crop and Food Research, Christchurch, New Zealand*, and Juliet A. Gerrard, *University of Canterbury, Christchurch, New Zealand*

How to obtain future titles on publication

A standing order plan is available for this series. A standing order will bring delivery of each new volume immediately upon publication. For further information, please write to:

Sales and Customer Care, Royal Society of Chemistry, Thomas Graham House, Science Park, Milton Road, Cambridge, CB4 0WF, UK
Telephone: +44(0) 1223 432360, E-mail: *sales@rsc.org*

RSC
FOOD
ANALYSIS
MONOGRAPHS

The Maillard Reaction

S.E. Fayle
Crop and Food Research, Christchurch, New Zealand

J.A. Gerrard
University of Canterbury, Christchurch, New Zealand

ROYAL SOCIETY OF CHEMISTRY

ISBN 0-85404-581-3

A catalogue record for this book is available from the British Library.

© The Royal Society of Chemistry 2002

Published by The Royal Society of Chemistry,
Thomas Graham House, Science Park, Milton Road, Cambridge CB4 0WF, UK
Registered Charity Number 207890

For further information see our web site at www.rsc.org

Typeset by Keytec Typesetting Ltd, Bridport, Dorset, DT6 3BE
Printed & bound in Great Britain by MPG Books Ltd, Bodmin, Cornwall, UK

Preface

The Maillard reaction, first described by Louis-Camille Maillard in 1912, is a general description of a complex series of reactions involving the reaction of free amino groups, such as amines, amino acids, peptides and proteins, with carbonyl compounds, particularly reducing sugars. Despite nearly ninety years of research on the Maillard reaction, its products have only recently started to be identified and the mechanistic pathways leading to their formation are only gradually being elucidated.

The Maillard reaction is particularly important in food systems where the products of the reaction can be responsible for the aroma, taste and appearance of foods. It can also cause deterioration of food during food storage and processing, resulting in a decrease in nutritional quality through the formation of anti-nutritional and toxic compounds, the destruction of essential amino acids and reduced digestibility of food proteins. Many researchers in the food area find themselves faced with a symptom of Maillard chemistry and are bewildered by the complexity of the reaction that they must try and comprehend in order to solve a particular food-processing problem.

This text is designed to provide a 'one-stop' shop for such researchers, providing a gentle introduction to the Maillard reaction and its many consequences, before focussing on the various methodologies employed to analyse the reaction in food. We hope to have provided an introduction to the field to those new in the area, as well as extend the analytical repertoire of those who are more experienced.

Siân Fayle
Juliet Gerrard
Christchurch, New Zealand, 2002

Contents

Acknowledgements

We would like to thank the many individuals who have helped us to bring this book to print.

First and foremost, a huge thank you to Nigel Larsen, for being the best boss on the planet. Thanks Nigel! Secondly, to the many individuals with whom we have discussed aspects of the subject and the text. Steve Elmore, Kevin Sutton, Nico Nibberin and John van der Graaf deserve particular mention for injecting wise counsel on specific chapters. Jackie Healy deserves a huge thank you for her technical wizardry. Matt Walters is eternally in our debt for his patience and skill, particularly during the production of the figures. Thirdly, to all our painstaking proof-readers – especially Susie Meade, Peter Steel, Antonia Miller and Ashley Sparrow – thank you for bedtime reading beyond the call of duty. Our errors, of course, remain our own.

Finally, the biggest thank you of all to those who put up with us. To the inhabitants of room 129, past and present, a big thank you for making the Maillard reaction fun. JAG would like to acknowledge the infinite patience of PJS, which has been truly tried and tested and say a huge thanks to Daniel and Lee for all the Saturday sleeps in and smiley faces. Siân would like to thank Susie Meade, Nik Solomon and Mike Kennedy for all the chocolate breaks, their support and especially for their friendship.

Abbreviations

1-DE	one-dimensional electrophoresis
2-DE	two-dimensional electrophoresis
AEDA	aroma extract dilution analysis
AFID	alkali flame ionisation detector
AMPS	ammonium persulfate
Bisacrylamide	N,N'-methylenebisacrylamide
BSA	bovine serum albumin
CE	capillary electrophoresis
CE–MS	capillary electrophoresis–mass spectrometry
CGE	capillary gel electrophoresis
CHAPS	3-[3-(chloroamidopropyl)dimethylammonio]-1-propanesulfonate
CI	chemical ionisation
CIEF	capillary isoelectric focusing
cmc	critical micellar concentration
CML	N-ε-carboxymethyllysine
CZE	capillary zone electrophoresis
dH$_2$O	distilled water
DTT	dithiothreitol
e-gram	electropherogram
ECD	electron capture detector
EOF	electro-osmotic flow
EI	electron impact
ELISA	enzyme-linked immunosorbent assay
ESI	electrospray ionisation
FAB	fast atom bombardment
FID	flame ionisation detector
FPD	flame photometric detector
FPLC	fast polymer liquid chromatography
FSCE	free solution capillary electrophoresis
GC	gas chromatography
GC–MS	gas chromatography–mass spectrometry
GCO	gas chromatography–olfactometry
GLC	gas–liquid chromatography

GSC	gas–solid chromatography
HMF	hydroxymethylfurfural (also known as 5-hydroxymethyl-2-furaldehyde and 5-(hydroxymethyl)furfural)
HPLC	high performance liquid chromatography
IE-HPLC	ion exchange high performance liquid chromatography
IEF	isoelectric focusing
IPG	immobilised pH gradient
kV	kilovolt
LC	liquid chromatography
LC–MS	liquid chromatography–mass spectrometry
MALDI	matrix-assisted laser desorption ionisation
MECC	micellar electrokinetic capillary chromatography
MEKC	micellar electrokinetic chromatography
MS	mass spectrometry
MS/MS	tandem mass spectrometry (also MS^2 or MS'')
m/z	mass:charge ratio
NMR	nuclear magnetic resonance
NPD	nitrogen–phosphorus detector
PAGE	polyacrylamide gel electrophoresis
PDMS	poly(dimethylsiloxane)
pI	isoelectric point
RNAse A	ribonuclease A
RP-HPLC	reverse-phase high performance liquid chromatography
SDS	sodium dodecyl sulfate
SDS–PAGE	sodium dodecyl sulfate–polyacrylamide gel electrophoresis
SIM	selected-ion monitoring
SPME	solid phase microextraction
TCA	trichloroacetic acid
TCD	thermal conductivity detector
TEMED	N,N,N',N'-tetramethylethylenediamine
TFA	trifluoroacetic acid
TID	thermionic ionisation detector
TLC	thin layer chromatography
TOF	time-of-flight mass analyser
Tris	N-tris(hydroxymethyl)methyl-2-aminoethanesulfonic acid
TMBD	N,N,N',N'-tetramethyl-1,3-butanediamine
UV	ultraviolet

CHAPTER 1

What is the Maillard Reaction?

1 A Little History

In 1912, Louis-Camille Maillard addressed the French Academy with a brief paper in which he described some recent experiments.[1] He had made a very simple observation: upon gently heating sugars and amino acids in water, a yellow-brown colour developed. The assembled learned Frenchmen may have been forgiven for not falling off their seats – even in 1912, this observation was surely not too surprising? However, Maillard was astute enough to realise that since biology is awash with sugars and amino acids, this reaction would have far-reaching implications.

Ninety years on, Maillard has acquired an impressive number of disciples. A Current Contents search of the keyword 'Maillard' yields over a thousand references for the period 1995–2000 alone. The consequences of Maillard chemistry are indeed extensive, and cut across many disciplines – the most noteworthy being food science and medicine. There cannot be many other fields in which the same fundamental chemical knowledge can be applied, for example, in both an interpretation of kidney disease and an understanding of why cooked onions taste pleasant. There are few reactions that attract such attention from organic chemists, food scientists and medics,[2] not to mention those working within the more obscure pockets of this widespread literature.[3,4] In this book, we will focus on the analysis of the Maillard reaction in food, but will occasionally draw on the chemical and medical literature where methodology has been developed that could usefully be adapted for food systems.

2 Chemical Definition

As with many reactions pertaining to biological systems, the precise chemical nature of the Maillard reaction is yet to be fully defined.[5] Early research, carried out by Maillard,[1] Amadori,[6] Kuhn and Weygand,[7,8] Simon and Kraus,[9] and Heyns *et al.*,[10] unravelled the earliest stages of the reaction which result in the so-called Amadori product (Figure 1.1). In a series of reversible reactions, the carbonyl moiety of the sugar molecule forms a Schiff base with a biological amine, typically an amino acid or the lysine residue of a protein. The resulting Schiff

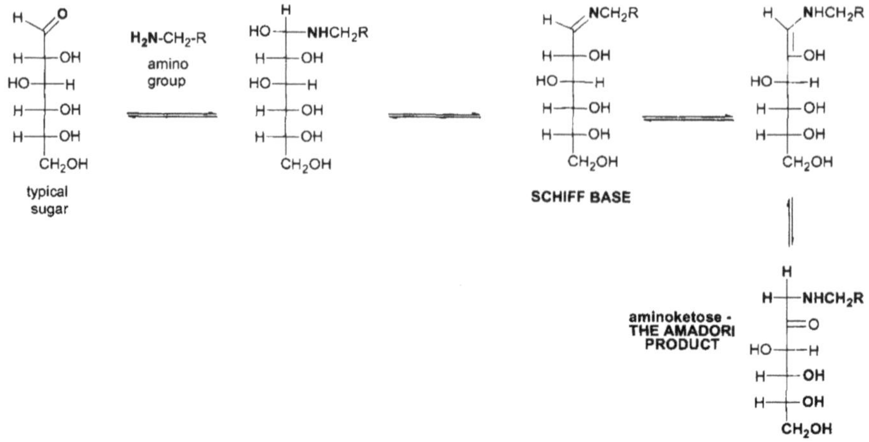

Figure 1.1 *The early stages of the Maillard reaction – from sugar and amino group to Amadori product – adapted from Ledl and Schleicher[5]*

base is labile and may undergo two sequential rearrangements, yielding a reasonably stable aminoketose – the Amadori product.

Further contributions to the early elucidation of the Maillard reaction pathways were made by Hodge.[11] Beyond the Amadori product, the situation starts to become rather more complex. The apparently simple reaction of an amino group with a monosaccharide leads to a huge variety of products in varying yields. Figure 1.2 illustrates some of the better-defined fates of the Amadori product. Each of these compounds is itself reactive and will participate in a vast array of possible reactions, the ratios of which vary according to the precise conditions that each molecule encounters.

Figures 1.1 and 1.2 depict the mere tip of an iceberg of known Maillard products. Furthermore, for each characterised compound, many other molecules remain poorly defined or unidentified. In addition to the large range of low molecular weight compounds that have been identified in model reactions of amino acids and sugars, substantial evidence points towards the formation of polymeric material during the later stages of the Maillard reaction. These high molecular weight, apparently heterogeneous, materials are often termed melanoidins and are notoriously difficult to characterise, although substantial progress has been made in recent years,[12-14] as will be seen in Chapter 9.

When the Maillard reaction takes place within a complex material, such as food, the reaction scheme is yet further complicated by the range of different compounds that may serve as reactants, including macromolecules. Thus, the analysis of the course of the Maillard reaction, under any particular set of conditions – for example during the drying of pasta[15] or within an eye lens[16] – represents a continuing challenge to scientists worldwide.

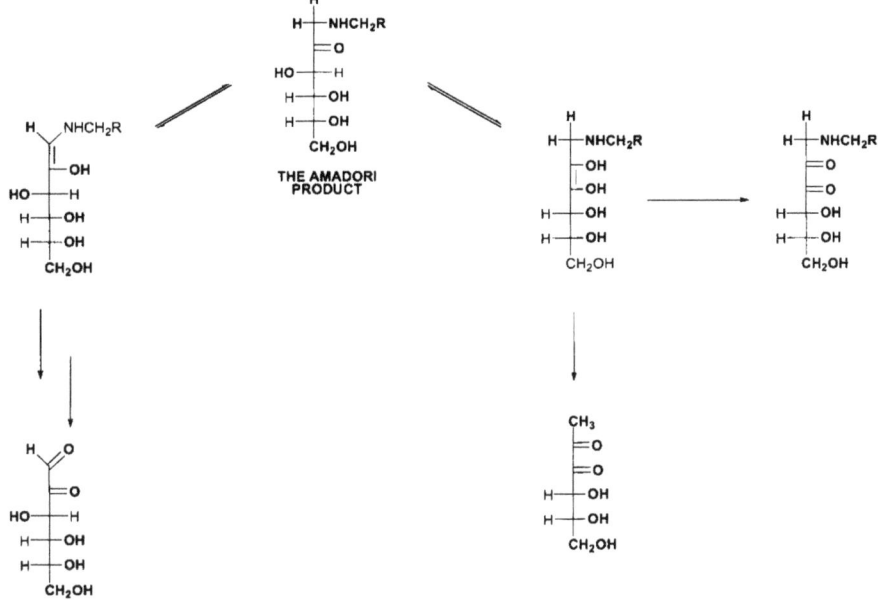

Figure 1.2 *Beyond the Amadori product -- adapted from Ledl and Schleicher*[5]

3 The Maillard Reaction *in situ*

As will be discussed in Chapter 2, much of our knowledge about the chemical mechanisms of the Maillard reaction has been derived from the analysis of so-called 'model systems' – chemical reactions designed to mimic the chemistry of foodstuffs or cells. What happens in processed food or inside cells? To answer this question, we need precise, accurate methods to measure the concentrations of Maillard products amongst an enormous number of potentially interfering substances. The situation is further muddied by the propensity of Maillard products to interconvert during the course of purification or measurement – an acute example of a fundamental problem encountered throughout any analysis of biological systems.

Biochemists and food chemists are faced with constantly shifting goalposts whenever they attempt to analyse the precise composition of a biological material. In any chemical study of a biological system, it is of paramount importance to understand the limits of the methodology being used, and the sets of conditions under which the measurements are relevant, which may, or may not, apply to the actual situation of interest. Nowhere are these constraints more apparent than in the analysis of the Maillard reaction in food systems – the subject of this book. Before embarking on an introduction to the specific problem of analysing the Maillard reaction in foods, the scene will be set with a brief overview of the limitations and challenges of analytical food chemistry in general.

4 Analytical Food Chemistry – Challenges and Limitations

Food chemistry is a vast science meeting many different types of need – from those interested in efficient and reliable tests to ensure quality assurance, to academic researchers seeking precise chemical descriptions of food processing.[17] The very different demands of these scientists lead to a disparate, and sometimes conflicting, body of literature in food chemistry. In a perfect world we would have a complete description of the chemistry of foods and a series of precise relationships between the molecular constitution of each foodstuff and its qualities. Unfortunately, we are a long way from knowing exactly which molecules, in which concentrations, are required to make a food taste or smell a particular way, or provide a particular mechanical property. More worryingly, perhaps, we remain largely ignorant of which molecules are toxic in our foods, and at what levels, and which combinations of ingredients may result in toxic products under given processing conditions – where pH, temperature and water activity may vary widely. Minute quantities of certain compounds may have an enormous effect on the flavour, aroma or carcinogenicity of a particular food[18-20] whilst others, in much higher concentrations, may have no effect at all.

Food chemistry is sandwiched between the conflicting approaches of its component disciplines. Faced with an analytical challenge in the carbohydrate arena, a chemist will start with a simple sugar molecule and work up; a food scientist, on the other hand, will start with a potato and work down; the food chemist must reconcile the two sets of results and create a model of the system consistent with both views. Progress in food chemistry is often made by striking a productive compromise between precise chemical definition and accurate measurement of a particular property of the food in question.

This 'compromise' approach often leads to working definitions that are confusing to those trained in more traditional, less applied sciences. There are many examples of classes of food chemicals that are defined by their method of preparation.[21] Thus, the substance in question is analysed by a reproducible method of analysis and accurately quantified. This leads to a 'standard method' of analysis for a substance that becomes so widely accepted that it is used to define its chemical nature! This conundrum is illustrated in Figure 1.3.

For example, the wheat gliadins constitutes a class of protein molecules that is defined as 'the fraction of wheat proteins that is soluble in 60–70% alcohol'.[22] This is an exasperating definition for someone seeking their precise chemical identity, but a very useful description for the food chemists who want to understand the variation in performance of flour derived from different cultivars of wheat, an active area of research around the world.[23-27]

Another example of the difference between the pure compounds that chemists describe and the substances commonly referred to by food scientists is illustrated by a common laboratory separation of the components of milk (Figure 1.4). In this preparation, lactose is prepared in a form that is readily recognisable to a chemist – it is a crystalline compound containing identical molecules of defined molecular structure. Casein, on the other hand, represents a heterogeneous mixture

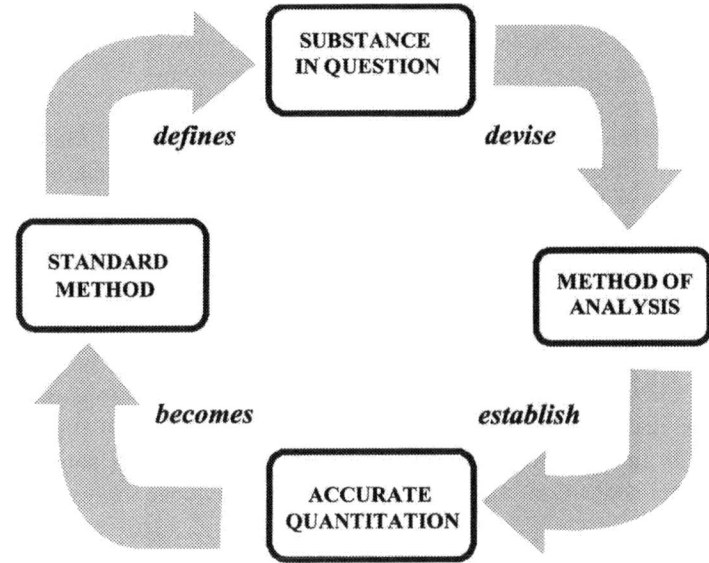

Figure 1.3 *Definition in food chemistry*

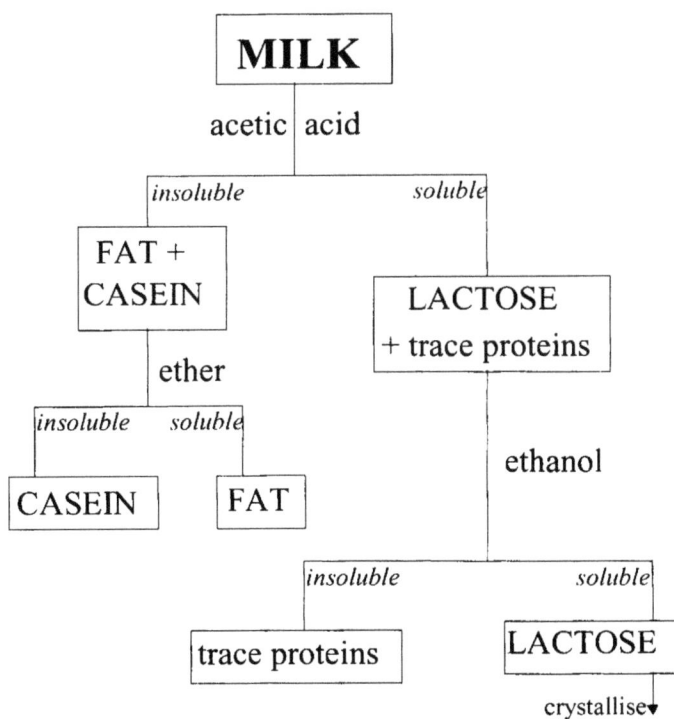

Figure 1.4 *A simple laboratory separation of the components of milk*

of proteins that is still being rigorously characterised as its components are separated and identified. The lack of detailed chemical knowledge of these food proteins has not prevented their useful separation from milk[28-31] in order to manipulate their properties in the vast international dairy industry.[32]

Thus, food chemists work with material that is not fully defined in a chemical sense. When trying to study a chemical reaction in food systems, therefore, they need to be aware that their starting material is not a pure substance, a fact which complicates the analysis of the products formed. They also need to appreciate how the reaction has affected the relevant properties of the food in question. Adding to this exacting challenge is the inherent variation in the precise make-up of any biological material – not only might the starting materials be heterogeneous and ill-defined, they may also vary according to the exact origins of the samples being analysed, the conditions under which they were grown, their age and the storage conditions under which they have been kept. All these factors must be borne in mind during any analytical food project, which constrains the generality of conclusions that can be drawn from any individual study.

5   Analysing the Maillard Reaction in Food

The conflicting demands of those requiring empirical, reproducible information and those endeavouring to piece together relationships between molecular structure and function in foods are confounded in situations where the products are numerous, changing rapidly or chemically undefined. Thus, to describe the analysis of the Maillard reaction in foods as challenging somewhat understates the task at hand. Simply to define the problem is often a project in itself, since we must identify precisely what it is we are looking for before we can even begin to work out how we might meaningfully measure it.

Several changes in the properties of foods have been attributed to the Maillard reaction. These include: changes in colour, particularly browning and, to a lesser extent, fluorescence; production of aroma and flavour compounds; production of bioactive compounds, both beneficial and toxic; loss of nutritional quality, especially of proteins; and changes in texture.[2] In order to correlate changes in each of these individual properties of a particular food with Maillard chemistry, quantitative methods are required for each property of interest. Quantifying sensory properties is no easy task, but once again, great strides have been made by those food chemists who manage to strike a balance between analytical stringency and experimental pragmatism. Successful measurement of some food properties often quoted as being altered by Maillard chemistry are considered in Chapter 2, with an illustrative case study in each instance.

6   Towards an Understanding of the Relationship between Food Qualities and Molecular Constitution

Many food qualities are affected by the Maillard reaction in foods during processing and storage and the potential rewards of deconvoluting this maze of reactions in our food are enormous. The food industry would, therefore, like to understand and harness the pathways of the Maillard reaction, enabling the improvement of existing products to meet the needs of industry and the consumer. Much effort has been invested in this process, as will be described in the pages of this book. The ultimate goal of all this research is to enable intervention and manipulation of this complex reaction pathway, in order to optimise the quality and safety of our food.

Much of our detailed understanding of the Maillard reaction has resulted from controlled model studies, carefully designed to mimic a particular situation encountered in the body or during food processing. Whether the conclusions reached in such studies truly represent reality is open to question. There remains a large gap in our knowledge that must be filled in order to relate the results of chemically defined experiments to those less defined studies that better mimic a given food processing situation. Bridging this gap is a challenge that relies on sophisticated analytical methodology, much of which is described in the pages that follow.

7 In a Nutshell

This book is not intended to provide a comprehensive review of the Maillard field, but instead presents carefully chosen examples from the recent literature to introduce the analytical methodology of the area. Case studies are used throughout, by way of introduction to the sort of analyses that can be carried out by food chemists in the Maillard field. We hope to inspire new experimental approaches for old systems and old experimental approaches for new ones.

Chapter 2 describes the many properties of food that may be altered by the Maillard reaction, and how we might begin to quantify them. Chapter 3 addresses the challenging area of how to extract Maillard reaction products from food, prior to analysis. The bulk of the book, Chapters 4 to 8, is organised according to analytical technique. For each chapter, a concise introduction to the scientific principles that underlie the technique is given, in order that a complete novice in the area may quickly appreciate the theory behind the experiment. The use of the specific technique in the Maillard arena is then surveyed.

A general criticism of the Maillard work carried out to date is the dependence on one or two instrumental techniques, and the lack of attempts to correlate results obtained from different studies. The final chapter describes some exceptions to this criticism, where the same problem has been approached from many angles, thereby yielding more valuable information. We hope to encourage other researchers to broaden their analytical repertoire in their quest to unravel their chosen corner of the Maillard reaction network, and relate their molecular findings to the overall properties of foods.

8   Further Reading

J. O'Brien, H.E. Nursten, M.J.C. Crabbe and J.M. Ames, *The Maillard Reaction in Foods and Medicine*, Royal Society of Chemistry, Cambridge, 1998.

R. Ikan, *The Maillard Reaction: Consequences for the Chemical and Life Sciences*, Wiley, Chichester, UK, 1996.

9   References

1. L.-C. Maillard, *C. R. Acad. Sci. Ser. 2*, 1912, **154**, 66.
2. J. O'Brien, H.E. Nursten, M.J.C. Crabbe and J.M. Ames, (eds.), *The Maillard Reaction in Foods and Medicine*, Royal Society of Chemistry, Cambridge, 1998.
3. C. Cole and J. Roberts, *Imaging Sci. J.*, 1997, **45**, 145.
4. A. McDonald, B. Fernandez, B. Kreber and F. Laytner, *Holzforschung*, 2000, **54**, 12.
5. F. Ledl and E. Schleicher, *Angew. Chem., Int. Ed. Engl.*, 1990, **29**, 565.
6. M. Amadori, *Atti R. Acad. naz. Lincei Mem. Cl. Sci. Fis. Mat. Nat.*, 1931, **13**, 72.
7. F. Weygand, *Chem. Ber.*, 1940, **73**, 1259.
8. R. Kuhn and F. Weygand, *Chem. Ber.*, 1937, **70**, 769.
9. H. Simon and A. Kraus, *Fortschr. Chem. Forsch*, 1970, **14**, 389.
10. K. Heyns, H. Paulsen, R. Eichstedt and M. Rolle, *Chem. Ber.*, 1957, **90**, 2039.
11. J.E. Hodge, *J. Agric. Food Chem.*, 1953, **1**, 928.
12. S.M. Rogacheva, M.J. Kuntcheva, T.D. Obretenov and G. Vernin in *The Maillard Reaction in Foods and Medicine*, J. O'Brien, H.E. Nursten, M.J.C. Crabbe and J.M. Ames, eds., Royal Society of Chemistry, Cambridge, 1998, 89.
13. T. Hofmann, W. Bors and K. Stettmaier, *J. Agric. Food Chem.*, 1999, **47**, 391.
14. L. Royle, R.G. Bailey and J.M. Ames, *Food Chem.*, 1998, **62**, 425.
15. R. Acquistucci, *Food Sci. Technol.*, 2000, **33**, 48.
16. S. Ramakrishnan, K.N. Sulochana, R. Punitham and K. Arunagiri, *Glycoconjugate J.*, 1996, **13**, 519.
17. O. Fenemma, *Food Chemistry*, 3rd ed., Marcel Dekker, New York, 1999.
18. W. Grosch and P. Schieberle, *Cereal Chem.*, 1997, **74**, 91.
19. M.S. Madruga and D.S. Mottram, *J. Sci. Food Agric.*, 1995, **68**, 305.
20. N. De Kimpe and M. Keppens, *J. Agric. Food Chem.*, 1996, **44**, 1515.
21. D.A.T. Southgate, *Determination of Food Carbohydrates*, 2nd ed., Elsevier Applied Science, London, 1991.
22. P.R. Shewry and A.S. Tatham, *J. Cer. Sci.*, 1997, **25**, 207.
23. H. Wieser, *Eur. J. Food Res. Technol.*, 2000, **211**, 262.
24. P.I. Torres, L. Vazquez-Moreno, A.I. Ledesma-Osuna and C. Medina-Rodriguez, *Cereal Chem.*, 2000, **77**, 702.
25. R. Kuktaite, E. Johansson and G. Juodeikiene, *Cereal Res. Comm.*, 2000, **28**, 195.
26. G. Branlard, *Ocl-Oleagineux Corps Gras Lipides*, 1999, **6**, 513.
27. B. Belderok, *Plant Foods Human Nutr.*, 2000, **55**, 1.
28. P. Punidadas and S.S.H. Rizvi, *Food Res. Int.*, 1998, **31**, 265.
29. S. Sachdeva and W. Buchheim, *Aust. J. Dairy Technol.*, 1997, **52**, 92.
30. Z. Jin, Y. Shukunobe and S. Taneya, *J. Jpn. Soc. Food Sci. Technol.*, 1997, **44**, 10.
31. K. Dinkov, A. Andreev and P. Panaiotov, *Milk Sci. Int.*, 1997, **52**, 127.
32. A. Andrews and J. Varley, *The Biochemistry of Milk Products*, Royal Society of Chemistry, Cambridge, 1994.

CHAPTER 2

Consequences of the Maillard Reaction in Food

1 Measuring Particular Qualities of Food

Many of the properties that make food appealing are, by their very nature, highly subjective qualities. The crispy coating of a roast chicken, the flaky texture of a freshly baked croissant or the subtle appearance of a delicate lemon sorbet, are all attributes that are recognised as desirable, but how one might quantify them is another matter. To place a quantitative score on 'chicken-skin crispiness' or 'pastry flakiness' is a challenge indeed. Sometimes this challenge is met with trained judges or consumer panels who rank particular foods for particular qualities; alternatively, instrumental methods may be developed. Each of these general approaches is subject to serious limitations, and to compensate, the most convincing studies of food properties often correlate instrumental investigations with consumer surveys.

As mentioned in Chapter 1, the Maillard reaction affects a whole host of sensory properties of food during storage and processing. If we are to understand these changes in molecular terms we must not only understand the chemistry taking place, but be able to link this chemistry to the quality of the food in question. This, in turn, requires reliable methodology to quantify each variable. In this chapter, a selection of food properties commonly cited as being affected by Maillard chemistry is discussed, with an emphasis on how these qualities are generally measured. This discussion provides a backdrop to the remainder of the book, which focuses more intently on how we might characterise the molecules responsible for these changes in our food.

2 The Maillard Reaction and Colour

The Maillard reaction is so commonly associated with colour formation that it is often dubbed 'the browning reaction' or 'non-enzymic browning',[1,2] the latter to distinguish this phenomenon from the equally common browning reactions caused by polyphenol oxidase and other enzymes.[3] Often, this browning is an integral part of the food processing of interest – and 'brownness' is the quality to which the consumer responds – and thus the property that the food chemist must measure. Elsewhere, browning has been used as a measurable symptom of the Maillard reaction, making the large, yet simplifying, assumption that the extent of browning provides a quantitative indicator of the extent of chemical reaction.

In many instances, Maillard browning is a highly desirable part of food processing. Without Maillard chemistry we would not have a dark bread crust or golden brown roast turkey, our cakes and pastries would be pale and anaemic, and we would lose the distinctive rich colour of French onion soup. In other cases, Maillard browning is detrimental to product quality if consumers find the browned product unappealing. Perhaps the best example of this latter phenomenon is in fruit preservation where Maillard browning, despite negligible effect on flavour, is extremely unattractive to customers.[4,5]

Measuring the extent of browning of a food, a food extract or an incubated model system is, at one level, trivially easy but, at another, dauntingly complex. At the simplest level, browning of a whole food can be monitored by visual inspection and comparison to standard samples, or by image analysis.[6,7] Food extracts and model systems, presuming that they can be made to form transparent, non-turbid solutions, can be monitored as they brown by following the absorbance at a chosen wavelength (typically[8] 420–460 nm). For those interested only in the colour of the sample, or those correlating molecular measurements with changes in colour, these methods are perfectly adequate. Case Study 2.1 outlines such a study, pertaining to the unwanted browning of orange juice.

CASE STUDY 2.1 – Browning in Orange Juice

Del Castillo *et al.*[9] have recently published a study in which they monitor the changes in amino acid composition in dehydrated orange juice due to Maillard browning. At 50 °C, 65% of the amino acids were lost after 14 days of storage. At 30 °C, the loss of amino acids was only 30%. The loss of amino acids was

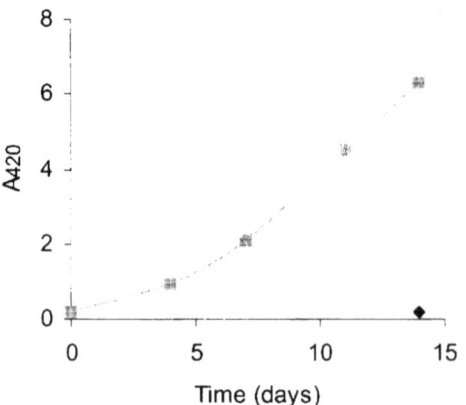

Figure 2.1   *Development of colour during storage at 50 (□) and 30 °C (♦) in dehydrated orange juice. Modified from del Castillo et al.[9]*

correlated with a drop in sucrose concentration, and fructose derivatives of the lost amino acids were detected. Browning was monitored by following the absorbance at 420 nm (Figure 2.1) and was found to occur in the later stages of the Maillard reaction, when these fructose derivatives were further degraded to heterogeneous pigments. The authors suggest, therefore, that these fructose-derived Amadori products be used as indicators of Maillard damage in dehydrated orange juice, since they can be detected before substantial deleterious colour has developed.

For those wishing to characterise the molecular changes that accompany browning, or follow the kinetics of specific Maillard events, simple photometric detection is severely limiting. From a chemist's standpoint, monitoring the rate of a process by measuring the collective absorbance of a complex mixture of compounds at an arbitrary wavelength is somewhat questionable. Such spectrophotometric measurements only yield meaningful kinetic information when one is measuring a single chromophore of known extinction coefficient – a condition that is clearly not met in a complex food system, or even a model system.

Research has been carried out to address this problem, and characterise the colour space and visual appearance of systems undergoing Maillard browning.[10] Case Study 2.2 shows the advances that have been made in deconvoluting the contribution of specific Maillard products to the total colour of browning solutions. Other researchers maintain that an increase in fluorescence provides a more reliable indicator of the extent of Maillard reaction.[11,12] These advances are encouraging, but have yet to find wide application in the analysis of actual food systems.

CASE STUDY 2.2 – Quantifying Colour in Single Maillard Products: The Colour Activity Concept

Hofmann[13,14] has developed the colour activity concept to address the surprising lack of knowledge as to the key chromophores that evoke Maillard browning. Maillard reaction products are ranked according to their colour impact using a colour dilution analysis method with synthetic reference compounds. This screening technique ranks the colorants on the basis of their relative colour intensities in solution, although it provides no insight into their

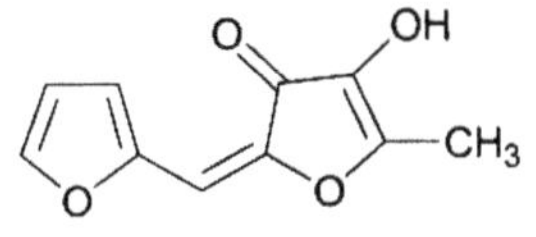

Figure 2.2 *2-[(2-Furyl)methylidene]-4-hydroxy-5-methyl-2H-furan-3-one – a reference compound of high colour activity[14]*

absolute contribution to the total colour of the browned Maillard mixture. Early results suggest that, despite the complexity of the network of browning reactions, large proportions of the total colour observed may be accounted for by a few key chromophores. An example of a significant chromophore in the model systems analysed is given in Figure 2.2.

3   The Maillard Reaction and Aroma and Flavour

In addition to the often desirable colour of foods that derives from the products of the Maillard reaction, many aroma and flavour compounds are also to be found amongst this complex network of molecules. The Maillard reaction plays a role in forming the distinctive flavours of many foods and beverages – among them chocolate,[15] coffee[16] and bakery products.[17] In fact, the aroma of most foods that are subjected to baking, roasting and grilling will contain Maillard reaction products. Almost invariably, hundreds of volatile compounds have been isolated from each food studied.[18]

The food industry has invested great effort in trying to create synthetic flavours and aromas by reconstituting combinations of these compounds, so called 'aromagrams', during processing. This approach has met with limited success, since the subtleties of aroma perception are many and varied, and the detection of these compounds, however sophisticated the separation system might be, must by necessity use a human nose as a detector.[19] Measurement of aroma and flavour is, by definition, a subjective process, and relating flavour to molecular structure is a very challenging field. Considerable progress has been made using the concept of 'odour activity', which is a parameter relating to the threshold concentration at which a compound must be present before its odour is detectable (see Case Study 2.3). Once these parameters are defined and quantified, changes in flavour and aroma during Maillard chemistry can be followed.

CASE STUDY 2.3 – The Odour of Fresh Strawberries

The chemical composition of the volatile compounds that constitute the aroma of strawberries has been extensively studied and over 360 molecules have been implicated in their distinctive flavour and aroma.[20] The most active flavour compounds have been identified by Schieberle and Hofmann[21] and a model strawberry juice has been reconstituted from these components, with their concentrations carefully balanced. Twelve of the most potent odorants were assembled in a matrix consisting of pectin and sugars and were found to produce a flavour profile that was very similar to genuine strawberry juice (Figure 2.3). The results suggest that a systematic approach to the location of key odorants using the odour activity concept, followed by careful reconstitution experiments, may allow great progress to be made in this arena.

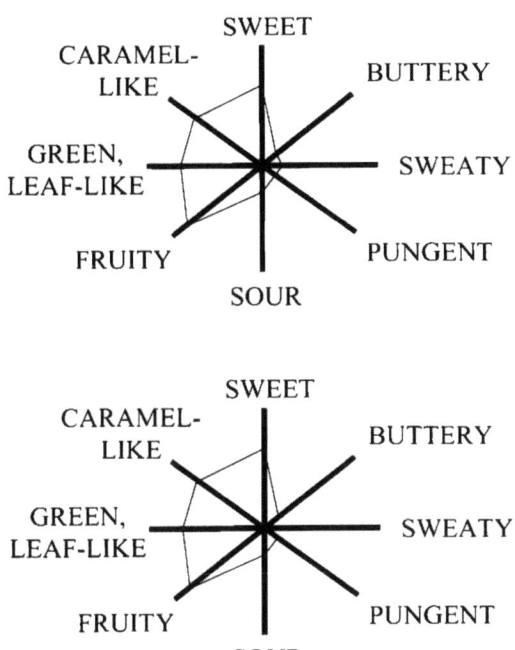

Figure 2.3 *Flavour profiles of the model strawberry juice (top) and freshly prepared strawberry juice (bottom), adapted from Schieberle and Hoffmann[21]*

Many of the key compounds isolated in such studies are likely to be reactive during Maillard chemistry. Thus, using the odour activity concept, we may start to understand the chemistry behind the distinctive changes in flavour and aroma that take place during the cooking of strawberries.

4   The Maillard Reaction and Texture

Although the effect of the Maillard reaction on colour, flavour and aroma is well documented, the effect on food texture has attracted less attention. However, increasing evidence suggests that within food systems the Maillard reaction can result in the crosslinking of proteins[22] and that protein crosslinking has profound effects on food texture.[23-27] This emerging area of research is likely to offer new methods for the manipulation of texture in processed foods.

Once again, the measurement of food texture can be a rather subjective process. Many instrumental methods are available to measure the mechanical properties of food,[28-31] but these need to be carefully related to the actual perception of texture by the consumer. An example is given in Case Study 2.4.

CASE STUDY 2.4 – Crosslinking Proteins to Manipulate Processed Food Texture

Hill and co-workers[32,33] have demonstrated that heating proteins and reducing sugars together reduces the solubility of the proteins, and that the proteins are covalently crosslinked by Maillard chemistry. Careful manipulation of the precise concentrations and heating profiles produced sugar–protein gels of specified breakstrength and clarity. They suggest that controlling protein crosslinking during food processing, for example in the barrel of an extruder, may allow manipulation of the properties of food texture in innovative and desirable ways.

---

5   The Maillard Reaction and Bioactivity

Amongst the quagmire of Maillard reaction products, there lurk many bioactive compounds, some of which may be beneficial, for example, with anti-oxidant properties,[34-38] and some toxic.[39,40] Although our understanding of the chemistry of formation of these compounds remains in its infancy, there are enormous potential benefits of understanding and controlling the Maillard reaction in such a way that beneficial compounds could be produced in the absence of unwanted molecules.

Measuring beneficial effects of Maillard products is not a simple process, since some sort of bioassay is required before we can be confident that any predicted effect is real. Ideally, a chemical test can be devised to screen products or reaction conditions before a bioassay is devised. Even then, there is no guarantee that the results of a bioassay will be relevant in the situation of interest. For example, anti-oxidants are purported to be amongst the many products of the Maillard reaction,[35] and such anti-oxidant activity can be detected in a chemical screen.[37,41-43] Once potentially useful compounds, or mixtures of compounds, are identified, they can be assessed for efficacy in food.[44] Beyond that, a bioassay using a cell line,[45] or an animal trial, typically with mice, might be devised to test the efficacy of the brew *in vivo*. Whether these results can be extrapolated to predict likely effects on human health is open to question.

In an analogous fashion, toxicity measurements in the literature are many and varied and subject to the same sort of constraints. Amongst the raft of commonly accepted toxicity tests[46-50] that are thought to be most relevant to human health is the Ames test.[51] A study of one form of toxicity attributed to Maillard chemistry, carcinogenicity, is described in Case Study 2.5.

CASE STUDY 2.5 – Carcinogens in Cooked Foods

Over fifty years ago it was demonstrated that extracts of cooked meat had carcinogenic properties, now known to be attributable to a series of heterocyclic amines formed through the Maillard reaction of beef, pork, lamb, chicken and fish muscle during cooking.[52] Careful analysis of such compounds (*e.g.* see Figure 2.4) in food that has been subjected to a variety of cooking methods suggests that particular cooking methods may heavily influence the likely production of these carcinogens. It may soon be possible for consumers to be advised of the safest way to cook certain foods, in order to maximise food safety.

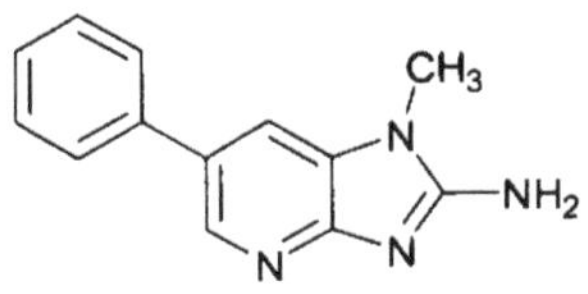

Figure 2.4 *2-Amino-1-methyl-6-phenylimidazo[4,5-b]pyridine – a carcinogen located in barbecued chicken*[52]

6 The Maillard Reaction and Nutrition

In addition to the effects of toxic, or anti-nutritional,[36] products of the Maillard reaction, the nutritional quality of foods may be damaged by the Maillard reaction during processing. This can occur for two main reasons. Firstly, aggregation of proteins on heating is known to decrease protein digestibility and is exacerbated by Maillard reactions.[53] Secondly, the Maillard reaction often leads to the loss of amino acids, especially lysine. Since lysine is commonly the limiting nutrient in a diet, its irreversible loss during Maillard chemistry can have a damaging effect on the overall nutritional value of a food.[36] This is of particular concern in areas such as the intensive livestock industry, where the precise nutritive value of a feed is of paramount importance for animal performance (see Case Study 2.6). A detailed understanding of the precise processing conditions required to maintain lysine, and other amino acids, may lead to more effective processes to generate feed of high nutritional value and reduce the need for feed supplements.

Once again, meaningful measurements of protein aggregation and amino acid availability are hard to devise. For example, there are several available methods for measuring the number of chemically available lysine residues in proteins,[36,53–57] but with any individual processed food sample these measurements do not necessarily agree. Most nutritional studies are only interested in the biological availability of the lysine residues, a parameter that can only be ascertained by feed trials that may or may not correlate to the chemical measurements. In an ideal world, simple chemical measurements can be used as an indicator and followed up with controlled feed trials (such as those described in Case Study 2.6).

CASE STUDY 2.6 – The Effect of Heat on Amino Acids in the Diet of Growing Pigs

Work by Barneveld *et al.*[58 61] has focussed on the effect of heat on amino acid availability in feed for growing pigs (Figure 2.5). Meticulous comparisons of

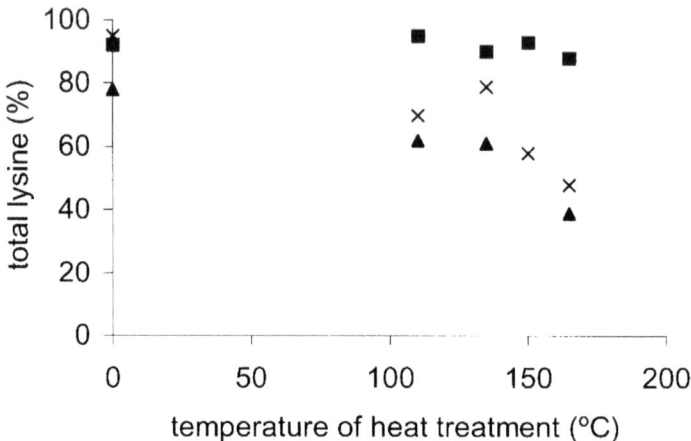

Figure 2.5 *The relationship between lysine digestibility (■), availability (×) and utilisation (▲) in raw field peas.[58 61] Heat treatment was for 15 minutes at the specified temperature and at low humidity. Adapted from Barneveld et al.[58 61]*

lysine digestibility, lysine availability and lysine utilisation showed that heat had little effect on ileal digestibility, but reduced the utilisation of ileal-digested lysine significantly, even at mild temperatures. This suggests that heating the feed, and thence increasing the rate of the Maillard reaction, renders lysine in a form that is chemically measurable and efficiently absorbed by the pig, but ineffectively utilised. This study serves to emphasise the difficulty in predicting nutritional value from simple chemical measurements, an area that requires large research efforts.

7   The Use of Model Systems

As we saw in Case Study 2.3, some of the changes to the properties of food, notably colour, flavour, aroma and the formation of bioactives, can be modelled in a simplified chemical brew – a model system. Typically, these contain simple mixtures of compounds, such as an individual amino acid and a sugar.[62-70] Other researchers have extended this concept to the use of proteins and sugars in model systems.[71-77] Once again, it is worth pausing to note that the simplification afforded in reducing the number of starting materials must be balanced against the loss of relevance to the system in question. However, considerable progress has been made in identifying Maillard reaction products through the use of model systems, and then locating these molecules in food. In the chapters that follow, the analysis of both food extracts and the products of model reactions by a variety of instrumental methods, will be considered. Prior to this, Chapter 3 will examine the extraction of Maillard reaction products from food.

8   In a Nutshell

The Maillard reaction has huge consequences on the properties of food, ranging from changes in visual appearance and flavour properties, to changes in the nutritional quality of the things we eat. Measurement of many of these food properties is a challenge in its own right. Correlating changes in the properties of food with the molecular changes caused by the Maillard reaction is an enormous area of research, which looks set to expand for the foreseeable future. A major hurdle to be overcome in order to achieve this goal is the extraction of molecules from the food in which they are generated. This will be considered in Chapter 3.

9   Further Reading

O. Fenemma, *Food Chemistry*, 3rd ed., Marcel Dekker, New York, 1999.

10   References

1. W.A.M. McMinn and T.R.A. Magee, *Food Bioprod. Process.*, 1997, **75**, 223.
2. C. Keller, B.L. Wedzicha, L.P. Leong and J. Berger, *Food Chem.*, 1999, **66**, 495.
3. M. Murata and S. Homma, *J. Jpn. Soc. Food Sci. Technol.*, 1998, **45**, 177.
4. S. Garza, A. Ibarz, J. Pagan and J. Giner, *Food Res. Int.*, 1999, **32**, 335.
5. A. Ibarz, J. Pagan and S. Garza, *J. Sci. Food Agric.*, 2000, **80**, 1162.
6. A. Ramirez-Jimenez, E. Guerra-Hernandez and B. Garcia-Villanova, *J. Agric. Food Chem.*, 2000, **48**, 4176.
7. P. Fernandez-Artigas, E. Guerra-Hernandez and B. Garcia-Villanova, *J. Agric. Food Chem.*, 1999, **47**, 2872.
8. J. Hutchings, *Food Colour and Appearance*, Blackie Academic and Professional, London, 1994.
9. M.D. del Castillo, N. Corzo, M.C. Polo, E. Pueyo and A. Olano, *J. Agric. Food Chem.*, 1998, **46**, 277.
10. D. MacDougall and M. Granov in *The Maillard Reaction in Foods and Medicine*, eds. J. O'Brien, H.E. Nursten, M.J.C. Crabbe and J.M. Ames, Royal Society of Chemistry, Cambridge, 1998, 464.
11. K.W. Lee, C. Simpson and B. Ortwerth, *Biochim. Biophys. Acta*, 1999, **1453**, 141.
12. F.J. Morales, C. Romero and S. Jimenez Perez, *Food Chem.*, 1996, **57**, 423.
13. O. Frank and T. Hofmann, *J. Agric. Food Chem.*, 2000, **48**, 6303.
14. T. Hofmann, *J. Agric. Food Chem.*, 1998, **46**, 3912.
15. M. Heinzler and K. Eichner, *Z. Lebensm.-Unters. Forsch.*, 1991, **192**, 24.
16. A.N. Wijewickreme and D.D. Kitts, *Food Chem. Toxicol.*, 1998, **36**, 543.
17. W. Grosch and P. Schieberle, *Cereal Chem.*, 1997, **74**, 91.

18. F. Ledl and E. Schleicher, *Angew. Chem., Int. Ed. Engl.*, 1990, **29**, 565.
19. T.E. Acree, *Anal. Chem.*, 1997, **69**, 170A.
20. A. Latrasse in *Volatile Compounds in Foods and Beverages*, ed. H. Maarse, Dekker, New York, 1991, 329.
21. P. Schieberle and T. Hofmann, *J. Agric. Food Chem.*, 1997, **45**, 227.
22. L.K. Peng, N. Ismail and A.M. Easa, *J. Food Sci. Technol.*, 2000, **37**, 188.
23. J.A. Gerrard, M.P. Newberry, M. Ross, A.J. Wilson, S.E. Fayle and S. Kavale, *J. Food Sci.*, 2000, **65**, 312.
24. J.A. Gerrard, S.E. Fayle, A.J. Wilson, M.P. Newberry, M. Ross and S. Kavale, *J. Food Sci.*, 1998, **63**, 472.
25. I.N.A. Ashie and T.C. Lanier, *J. Food Sci.*, 1999, **64**, 704.
26. N. Seki, H. Nozawa and S.W. Ni, *Fish Sci.*, 1998, **64**, 959.
27. H.Y. Ting, S. Ishizaki and M. Tanaka, *J. Muscle Foods*, 1999, **10**, 279.
28. H.A. Bremner, *Crit. Rev. Food Sci. Nutr.*, 2000, **40**, 83.
29. S.S. Narine and A.G. Marangoni, *Food Res. Int.*, 1999, **32**, 227.
30. M. Peleg, *Food Sci. Technol. Int.*, 1997, **3**, 227.
31. M. Pons and S.M. Fiszman, *J. Texture Stud.*, 1996, **27**, 597.
32. S.E. Hill and A.M. Easa in *The Maillard Reaction in Foods and Medicine*, eds. J. O'Brien, H.E. Nursten, M.J.C. Crabbe and J.M. Ames, Royal Society of Chemistry, Cambridge, 1998, 133.
33. Z.H. Mohammed, S.E. Hill and J.R. Mitchell, *J. Food Sci.*, 2000, **65**, 221.
34. M. Anese, L. Manzocco, M.C. Nicoli and C.R. Lerici, *J. Sci. Food Agric.*, 1999, **79**, 750.
35. P. Bersuder, M. Hole and G. Smith, *J. Am. Oil Chem. Soc.*, 1998, **75**, 181.
36. M. Friedman, *J. Agric. Food Chem.*, 1996, **44**, 631.
37. D. Mastrocola and M. Munari, *J. Agric. Food Chem.*, 2000, **48**, 3555.
38. Y. Yoshimura, T. Iijima, T. Watanabe and H. Nakazawa, *J. Agric. Food Chem.*, 1997, **45**, 4106.
39. M. Alonso and J. Zapico, *J. Food Biochem.*, 1995, **18**, 393.
40. Z. Balogh, J.I. Gray, E.A. Gomaa and A.M. Booren, *Food Chem. Toxicol.*, 2000, **38**, 395.
41. M. Pischetsrieder, F. Rinaldi, U. Gross and T. Severin, *J. Agric. Food Chem.*, 1998, **46**, 2945.
42. A.N. Wijewickreme, Z. Krejpcio and D.D. Kitts, *J. Food Sci.*, 1999, **64**, 457.
43. D. Mastrocola, M. Munari, M. Cioroi and C. R. Lerici, *J. Sci. Food Agric.*, 2000, **80**, 684.
44. F. Bressa, N. Tesson, M. Dallarosa, A. Sensidoni and F. Tubaro, *J. Agric. Food Chem.*, 1996, **44**, 692.
45. N. Ide, B.H.S. Lau, K. Ryu, H. Matsuura and Y. Itakura, *J. Nutr. Biochem.*, 1999, **10**, 372.
46. K. Bottrill, *ATLA-Altern. Lab. Anim.*, 1998, **26**, 421.
47. G.J. Harry, M. Billingsley, A. Bruinink, I.L. Campbell, W. Classen, D.C. Dorman, C. Galli, D. Ray, R.A. Smith and H.A. Tilson, *Environ. Health Perspect.*, 1998, **106**, 131.
48. M. deBel, L. Stokes, J. Upton and J. Watts, *Water Sci. Technol.*, 1996, **33**, 289.
49. J.A. Swenberg and R.O. Beauchamp, *Crit. Rev. Toxicol.*, 1997, **27**, 253.
50. A. Kilroy and N.F. Gray, *Biol. Rev. Cambridge Philosophic. Soc.*, 1995, **70**, 243.
51. K. Mortelmans and E. Zeiger, *Mutat. Res.-Fundam. Mol. Mech. Mutagen.*, 2000, **455**, 29.
52. J. Felton and M. Knize in *The Maillard Reaction in Foods and Medicine*, eds. J. O'Brien, H.E. Nursten, M.J.C. Crabbe and J.M. Ames, Royal Society of Chemistry, Cambridge, 1998, 11.
53. R.F. Hurrell and P.A. Finot in *Digestibility and Amino Acid Availability in Cereals and Oilseeds*, eds. J.W. Finley and D.T. Hopkins, AAC Inc, St Paul, Minnesota, 1985, 233.
54. R.F. Hurrell and K.J. Carpenter, *Prog. Food Nutr. Sci.*, 1981, **5**, 159.

55. M. Friedman, J. Pang and G.A. Smith, *J. Food Sci.*, 1984, **49**, 10.
56. C. Bertrand-Harb, M.-G. Nicolas, M. Dalgalarrondo and J.-M. Chobert, *Sciences des Aliments*, 1993, **13**, 577.
57. P.J. Moughan and S.M. Rutherford, *J. Agric. Food Chem.*, 1996, **44**, 2202.
58. R.J. Barneveld, E.S. Batterham and B.W. Norton, *Br. J. Nutr.*, 1994, **72**, 243.
59. R.J. Barneveld, E.S. Batterham, D.C. Skingle and B.W. Norton, *Br. J. Nutr.*, 1994, **73**, 259.
60. R.J. Barneveld, E.S. Batterham and B.W. Norton, *Br. J. Nutr.*, 1994, **72**, 221.
61. R.J. Barneveld, E.S. Batterham and B.W. Norton, *Br. J. Nutr.*, 1994, **72**, 357.
62. L.P. Leong and B.L. Wedzicha, *Food Chem.*, 2000, **68**, 21.
63. M. Bristow and N.S. Isaacs, *J. Chem. Soc., Perkin Trans. 2*, 1999, **10**, 2213.
64. O. Mandin, S.C. Duckham and J.M. Ames, *J. Agric. Food Chem.*, 1999, **47**, 2355.
65. A. Sensidoni, D. Peressini and C.M. Pollini, *J. Sci. Food Agric.*, 1999, **79**, 317.
66. J.M. Ames, R.G. Bailey and J. Mann, *J. Agric. Food Chem.*, 1999, **47**, 438.
67. T. Hofmann and S. Heuberger, *Food Res. Technol.*, 1999, **208**, 17.
68. Y. Al-Abed, P. Ulrich, A. Kapurniotu, E. Lolis and R. Bucala, *Biorg. Med. Chem. Lett.*, 1995, **5**, 2929.
69. Y. Al-Abed, D. Callaway, A. Kapurniotu, T. Holak, W. Voelter and R. Bucala, *Polish J. Chem.*, 1999, **73**, 117.
70. A. D'Agostina, M. Negroni and A. Arnoldi, *J. Agric. Food Chem.*, 1998, **46**, 2554.
71. V. Fogliano, S.M. Monti, T. Musella, G. Randazzo and A. Ritieni, *Food Chem.*, 1999, **66**, 293.
72. R.L. Bunde, E.J. Jarvi and J.J. Rosentreter, *Talanta*, 2000, **51**, 159.
73. L. Pellegrino, I. De Noni and S. Cattaneo, *Nahrung*, 2000, **44**, 193.
74. R. Stockmann and R.K. Weerakkody, *Aust. J. Dairy Technol.*, 2000, **55**, 108.
75. D.G. Dyer, J.A. Blackledge, S.R. Thorpe and J.W. Baynes, *J. Biol. Chem*, 1991, **266**, 11654.
76. J.A. Gerrard, S.E. Fayle and K.H. Sutton, *J. Agric. Food Chem.*, 1999, **47**, 1183.
77. S.E. Fayle, J.A. Gerrard, L. Simmons, S.J. Meade, E.A. Reid and A.C. Johnston, *Food Chem.*, 2000, **70**, 193.

CHAPTER 3

Extraction of Maillard Reaction Products from Food

1 Introduction

In Chapter 2, the various manifestations of the Maillard reaction in food were described, with an indication of how changes in food properties are often monitored. In order to correlate the changes in food during processing with particular molecular reactions, we need to extract the reacted molecules from the food of interest and identify them. The remaining chapters of this book are largely devoted to the instrumental techniques available to separate and characterise the complex mixtures of molecules that arise in food systems during Maillard chemistry. In this chapter, we take a general look at how such molecules might be extracted from various foodstuffs, prior to analysis.

As with many analytical challenges, the sampling method may strongly influence the results of the analysis itself. In many ways, therefore, the extraction method chosen may be considered the most important part of the experimental design. Before embarking on an extraction of Maillard products from a particular food, the purpose of the analysis must be clearly defined and two major issues addressed. Firstly, is information sought about a particular molecule, or set of molecules, only? Or is information required about any or all of the potential molecules in the food that might have reacted during processing? Secondly, does the chosen method of analysis impose any constraints on the nature of the extraction? In the next two sections, each of these issues will be addressed in turn.

2 What Information is Required?

The nature of the question being posed during any Maillard reaction study should heavily influence the type of extraction method selected. In some cases, the formation of a particular molecule, or biomarker, might be the subject of interest. This approach is prevalent in the study of the Maillard reaction in medicine, where an increasing number of studies cite the monitoring of biomarkers as accurate methods of monitoring the extent of Maillard damage in the body. Examples include the many published studies reporting the presence of N-ε-

carboxymethyllysine (CML)[1-3] or pentosidine[4,5] as diagnostic indicators of tissue ageing or damage. Such approaches are also finding application in the study of the Maillard reaction in food. For example, if a particular Maillard product is toxic, and its formation is being monitored as a function of reaction conditions, the concentration of this molecule is the only piece of information sought, *e.g.* hydroxymethylfurfural (HMF), which has mutagenic potential.[6] Other examples include the need to identify and quantify specific off-flavours in foods.[7]

In many ways, developing a method to monitor the concentration of a single chemical compound, or group of compounds, is substantially easier than monitoring a complex mixture in its entirety. There are two main approaches: the molecule of choice may have a distinctive characteristic that allows it to be quantified without substantial purification; or the molecule may lend itself to a selective method of extraction, relatively free from contamination. In the first scenario, the distinctive characteristic may be as simple as a unique retention time in a particular chromatographic method,[8,9] or it may be something more sophisticated, such as the ability to be detected by a specific antibody in an immunoassay. The latter technique shows great potential for the monitoring of single Maillard products of interest and often requires minimal sample extraction and preparation. This approach will be further explored in Chapter 9.

In circumstances where the products of interest are, as yet, unknown, or when information about the complete spectrum of Maillard reactivity is required, extraction methods must aim to be as complete as possible, so that all relevant compounds are extracted for analysis. At this point, the investigator must compromise between being confident that the complete complement of the molecules of interest have been extracted, and concerned that the molecules may have changed during the extraction process, thereby creating artefacts. As a first approximation, most molecules will be more successfully extracted under harsh conditions, for example when the sample is heated. However, such extreme conditions are also more likely to create artefactual results. Careful controls can minimise the probability that artefacts are created during extraction, but this possibility can never be eliminated.

On the other hand, where artefact formation is controlled and understood, it may be used diagnostically to infer the presence of a particular molecule in the sample. This is often the case with methods that are employed to extract compounds for analysis by gas chromatography (Section 5). Another example is the harsh, but reproducible, hydrolysis of proteins to their constituent amino acids, prior to standard analysis by high pressure liquid chromatography (HPLC). This technique is widely used to detect proteins that have been modified by the Maillard reaction,[10] although the degree of artefact formation by the very acidic conditions of the protein digest is not commonly acknowledged.[11]

3 Instrumental Constraints

To a large extent, sample preparation is constrained by the instrument that will be used for analysis. For liquid chromatography, the subject of Chapter 5, the sample must be soluble in a solvent that is compatible with the column to be used for the separation, and free from contaminants that may damage the column. For example, before food proteins in pastry can be analysed by size exclusion chromatography, all fat must be removed from the sample. This can be achieved *via* Soxhlet extraction (see Section 5) under conditions that do not affect the composition of the proteins under study.[12]

The solvent in which the sample is to be analysed may not always be the most appropriate solvent with which to extract the molecule from a food sample. If this is the case, then the extraction may take place in the solvent of choice, after which the extraction solvent is removed by evaporation, and the sample re-suspended as required. However, in all extraction and manipulation procedures, the investigator should remember that each additional step of the protocol introduces added potential for artefact formation. This is particularly true when a solvent is removed from a dilute solution containing reactive molecules, as the higher concentrations of molecules are more prone to react.

Most of the techniques described in this book – notably liquid chromatography, mass spectrometry, electrophoresis and capillary electrophoresis – rely on the molecules being soluble in appropriate solvents. The nature of the extraction procedure is often very specific to the particular molecules, and food, in question. However, a brief treatment of some general considerations for the extraction of soluble Maillard products from food is given in Section 4. A notable exception to this requirement for solubility is gas chromatography (Chapter 4), which analyses molecules in the gas phase. Since a host of Maillard chemistry has been unveiled using this important technique, and extraction of molecules for gas chromato-graphic analysis imposes very particular requirements for the researcher, this is discussed in some detail in Section 5.

4   Sampling Techniques for Soluble Molecules

Certain foods, particularly beverages and other liquid products, may be amenable to direct analysis by liquid chromatography, for example soya sauce,[13] beer[14] and fruit juice.[15] However, the majority of foods do not yield so readily to detailed analysis. Typical procedures for extracting molecules from foodstuffs include homogenising the sample with an appropriate solvent, and centrifuging the resulting slurry to remove remaining solid matter. The supernatant liquid constitutes a crude extract of Maillard reaction products, along with many other molecules that were present in the food. Depending on the nature of the food in question, and the information sought (Section 2), this crude extract may be ready for direct analysis. For example, in the study of the Maillard reaction of gluten and glucose, the reacted molecules can be extracted with methanol.[16] These simple extractions allow direct comparison with the analysis of model systems, typically of a sugar and an amino acid, where direct analysis of the resulting mixture is often possible, or where solvent extraction is the only sample preparation that is required.[17-22]

Some foods require the specific removal of substances that interfere with the analysis. For example, in the analysis of Maillard reaction products in beetroot, a method has been developed to reduce the background level of sucrose, prior to analysis.[23] Alternately, a series of steps may be required to reduce the molecules being analysed to a manageable number.[24]

5 Sampling Techniques for Volatile Molecules

The analysis of volatile molecules is somewhat more challenging than the analysis of those that are non-volatile, because, by definition, they tend to escape as a gas. Some analyses of volatile compounds adopt an approach very similar to the methods described in Section 4 – extraction of the desired compounds into an appropriate solvent. This is very effective if the food is aqueous-based and the volatiles of interest dissolve in an organic solvent. An excellent recent example of this approach is a series of Maillard reaction products characterised from the honey of the Australian blue gum, which were extracted using ethyl acetate before being subjected to gas chromatographic (GC) analysis.[25]

Where simple solvent extraction is not sufficient, many different techniques are available for the specific isolation of volatile Maillard reaction products, including distillation, extraction, pyrolysis and headspace analysis.[7,26–28] Each technique is appropriate for the preparation of samples for GC analysis, and each has inherent advantages and disadvantages. All tend to be limited by the potential destruction of the volatile target compounds and/or the production of volatile artefacts. The number of compounds that can be analysed by GC ranges beyond those that are volatile at room temperature: heating the sample, or applying a vacuum, increases the volatility of other compounds, as does derivatisation. Even more so than with other instrumental techniques, the results obtained from GC analysis are heavily influenced by the choice of extraction technique employed, as well as the nature and stability of the compounds being studied.

Derivatisation

Not all compounds are suitable for analysis by GC, due to either their thermal instability or lack of volatility. In these instances, chemical derivatisation may enable GC analysis by creating a stable and/or volatile derivative. The success of characterising a molecule after derivatisation relies on an intimate knowledge of the chemistry involved, and how this relates to the structures of the molecules that are being analysed. Compounds that often require derivatisation include those containing hydroxyl, carboxylic acid or amino functional groups.[29] Each of these functional groups contributes to the hydrogen-bonding ability of the compound. As hydrogen bonding reduces the volatility of a compound, derivatisation that removes or masks the hydrogen-bonding group increases the volatility of the compound and facilitates its analysis by GC (see Figure 3.1).

Derivatisation techniques can also be used to improve the resolution or selectivity of a GC technique. For example, a higher detector response may be achieved by incorporating a particular functional group in the target compounds,

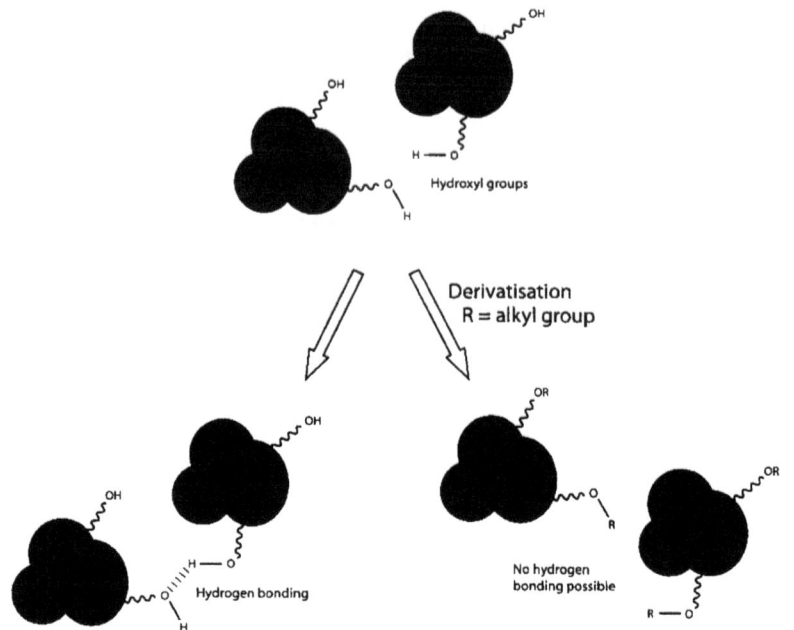

Figure 3.1   *Derivatisation may increase the stability or volatility of a molecule by reducing hydrogen bonding*

such as derivatisation with a halogenated functional group for use with an electron capture detector (ECD) (see Chapter 4). Other examples of standard methods for derivatisation include silylation, alkylation or acylation.[30,31]

Distillation

Distillation methods, either at atmospheric pressure or under reduced pressure, have been two of the more popular methods of volatile extraction in the food arena.[32,33] Distillation has the advantage of isolating only the volatile components of a sample, avoiding contamination by non-volatile compounds that may interfere with the GC analysis. Steam distillation may also be employed, although this has the disadvantage of producing dilute aqueous solutions of the extracted compounds, necessitating concentration of the sample prior to analysis.

Solid Phase Extraction

Volatiles may be successfully extracted directly from a solid sample.[32] One of the most well known methods of solid phase extraction is through the use of the Soxhlet extraction apparatus. This is an aggressive method that involves the direct removal of soluble components from a solid sample by repeated extraction from the solid with a refluxing solvent, chosen for its selectivity and boiling point. Prior to analysis, the volatiles are concentrated by the removal of the solvent, often under reduced pressure. This concentration step may result in the loss of some important components, resulting in an incomplete spectrum of molecules being recorded. Contamination of the sample by the incomplete removal of the solvent, or by the simultaneous extraction of non-volatile food components, may also lead to artefactual results.

Likens–Nickerson Simultaneous Steam Distillation–Extraction

The combination of distillation and extraction methods is often used for the analysis of Maillard reaction products.[34-37] This is usually achieved with the use of the Likens–Nickerson simultaneous steam distillation–extraction apparatus, which has been a standard method of isolating molecules from food for a long time (see Case Study 3.1).[38,39] Advantages of this method include: the requirement of low solvent volumes, thereby minimising cost; the ability to rapidly concentrate the volatiles in a single step, producing a concentrated extract that can be used for multiple analyses; and finally, with the use of reduced pressure, the ability to minimise thermal degradation. Although reducing the pressure decreases the temperature required for the volatilisation of the target compounds, it also leads to solvent losses. This can, however, be minimised by the use of a cold trap.

Pyrolysis

As many synthetic polymers and macromolecules, such as the large melanoidins generated by the Maillard reaction (Chapter 1), are involatile, they are not suitable for direct analysis by GC. An alternative, therefore, is pyrolysis of the sample, one of the most common sampling methods utilised for the gas phase study of Maillard reaction products.[40-45] This involves the rapid and controlled heating of the sample to a predetermined temperature, ranging anywhere from 400 °C to 1000 °C. Any volatile fragment molecules will be carried onto the GC column where they are separated.[46,47] The chromatographic pattern of the fragments provides a fingerprint that is characteristic of the sample and can be used for its identification.

Headspace Analysis

Headspace analysis is the direct analysis of the vapour that collects above the sample, and minimises the formation of artefacts commonly associated with derivatisation. A simpler, but less complete, chromatogram is produced using headspace techniques than when solvent extraction or distillation methods are used,[48] due to the more aggressive nature of the latter sampling methods. The simplest technique available for the analysis of headspace volatiles is the static method.

Static headspace analysis involves the equilibration of a sample in a sealed container, such that the components of the sample partition between the food matrix itself and the vapour phase immediately surrounding it. An aliquot of headspace gas can then be withdrawn and injected directly into the GC. Although this method is simple, rapid and reliable, it does not allow the analysis of components that have higher boiling points, which may be significant for the odour properties of a foodstuff, particularly when analysing foods that are typically served heated. Also, as the volatility of a compound is affected by its interaction with other non-volatile components of the reaction mixture, such as lipids, proteins and carbohydrates, static headspace analysis suffers from relatively low sensitivity.

Various methods have been developed that enrich the headspace gas, prior to collection, further improving the sensitivity of the analysis. These include the steam distillation–solvent extraction method described previously, cryotrapping and the 'purge and trap' method of volatile collection. Cryotrapping involves passing an inert gas through the sample and into a cold trap where the volatiles are deposited. Unfortunately, large volumes of water also condense in the trap necessitating the thawing of the trap and subsequent extraction of the volatiles by solvent extraction. This method is time-consuming and has a high potential for artefact formation.

The more favoured method of enriching headspace gases is adsorbent trapping. As for cryotrapping, an inert gas, often oxygen-free nitrogen, is passed through the sample and the gas stream carries the volatiles into an adsorption trap, as illustrated in Figure 3.2. Various adsorbents can be used, including activated charcoal, molecular sieves and the popular porous polymers, such as Tenax.[49] Tenax is a thermally stable material with a high absorptive capacity toward volatile compounds and a low affinity for water vapour, and so is highly suited to the collection of flavour volatiles.[50] After collection, the trap is disconnected and purged with purified nitrogen to remove any condensed water. The volatiles can be stored in the porous polymer traps for up to two weeks prior to analysis,[51] after which they can be recovered by solvent extraction or distillation. By far the easiest method of analysis, however, is the thermal desorption of the volatiles

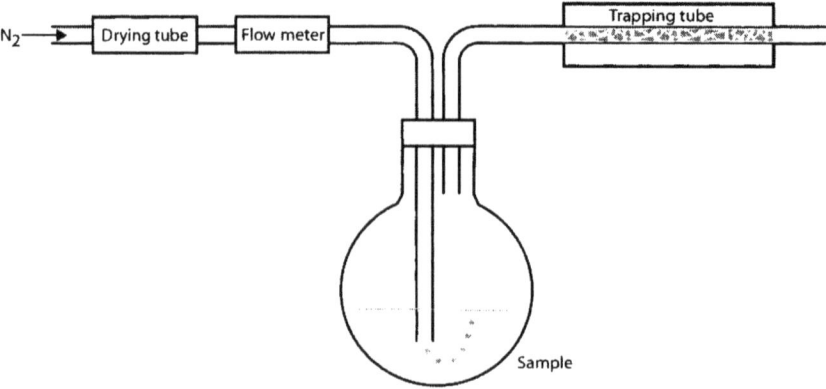

Figure 3.2   *Schematic of a typical purge and trap apparatus*

directly onto the GC column. Thermal desorption involves the rapid heating of the trapping tube, at temperatures ranging from 50 °C to 350 °C, leading to the transfer of the desorbed volatiles directly into the injection port of the GC. The purge and trap procedure is widely used by researchers studying volatile products of the Maillard reaction, as exemplified in Case Study 3.1.[52-54]

CASE STUDY 3.1 – Flavour Volatiles of Extruded Products

Extrusion cooking has become an extremely popular processing technique in the food industry, where it is used to create food products such as breakfast cereals, expanded snacks, confectionery and petfoods. It is a highly versatile process as it can complete many functions in a single operation, including mixing, cooking, foaming, puffing and drying.[53]

Extrusion is a high shear, high pressure, short time process that involves the feeding of raw materials, such as corn flour, into a heated barrel. The materials are mixed and transported along the barrel by either one or two turning screws. When the material reaches the end of the barrel, it is squeezed out through a small opening, known as a die, where it experiences a sudden drop in pressure. This causes the water within the product to immediately vaporise, resulting in the characteristic puffed product.[54]

The Maillard reaction plays an important role in the development of the flavour, aroma and colour profiles of extruded foods, since the low moisture and high temperature conditions of the extrusion process favour Maillard chemistry. However, one of the problems associated with extrusion is the general lack of flavour of extruded products. This is attributed to the loss of volatile flavour compounds as the food material exits the extruder, due to a process analogous to steam distillation, as the water contained in the extrudate vaporises. Additionally, there is insufficient time available for the generation of new flavour volatiles.[55] This problem can be addressed by the addition of flavour compounds either before or after extrusion. Flavour blends added after extrusion can lead to the uneven distribution of flavour, often resulting in a bland central core. The addition of flavourings prior to extrusion however is costly, since large quantities are needed to counter losses due to steam distillation, and may result in the generation of off-flavours. An alternative is to protect the flavour compounds by encapsulating them in complex polysaccharides, proteins or hydrocolloids prior to adding to the raw materials. Even this method is thought to require high initial concentrations to ensure sufficient flavour in the final product.[56]

The loss of flavour volatiles has been measured by Nair *et al.*[57] Their study involved the collection of thermally generated volatiles, released at the die during processing, which were then compared with those recovered from the corresponding solid extrudate matrix.

A high amylose corn flour was extruded and the volatiles were collected at the die using an apparatus consisting of a series of cold traps, shown in Figure 3.3. As this apparatus also collects most of the condensed steam, the cold traps

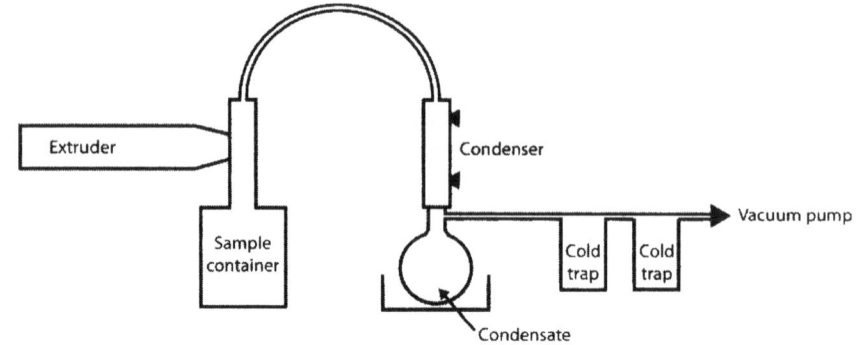

Figure 3.3   *Apparatus for collecting volatiles at the extruder die. Adapted from Nair et al.[57]*

(A)

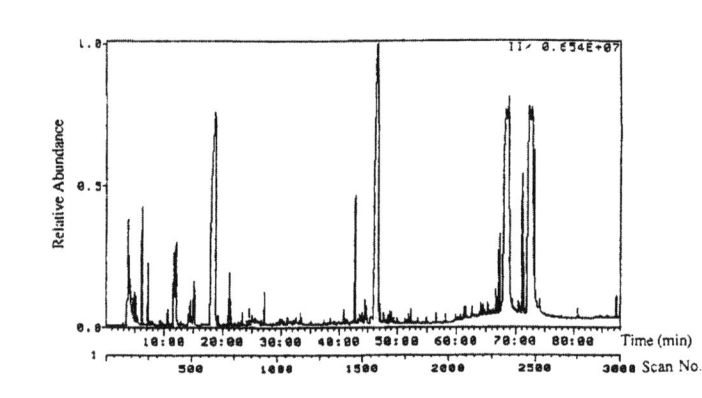

(B)

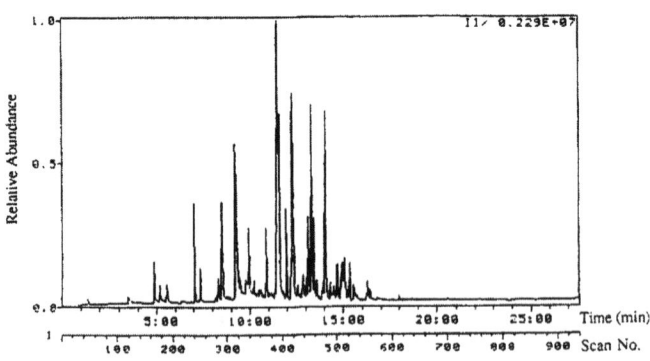

Figure 3.4   *Volatile profiles of the compounds (A) released at the die and (B) recovered from extrudates*
(Reproduced with permission[57])

were each washed with methylene chloride immediately after collection and combined with the water condensate. The sample was extracted by steam distillation–extraction, using the Likens–Nickerson method, concentrated and analysed by gas chromatography–mass spectrometry (GC–MS).

The volatiles isolated directly from the extruded product were collected in an adsorbent tube, containing an equal quantity of Tenax TA and Carbotrap, by the purge and trap method, and were analysed by thermal desorption-GC–MS.

The volatile profiles are shown in Figure 3.4. A total of 91 compounds were identified among the volatile compounds collected at the die, compared with 56 from the corresponding extrudate sample. It was observed that the types of volatile compounds collected at the die differed significantly from those collected from the extrudate, although the authors suggested that the different sampling methodologies made it difficult to compare the two fractions directly. Comparison of the identified compounds indicated that 16 were common to both samples, most of which are known to be important flavour contributors, confirming that there is a significant loss of flavour volatiles at the extruder die.

A further sampling technique, known as solid phase microextraction (SPME), has been developed by Pawliszyn and co-workers[58,59] and can be used to extract and concentrate a wide range of volatile or semi-volatile organic compounds.[60] It is a solvent-free sampling technique that uses a fused silica fibre, coated with an adsorbent material such as poly(dimethylsiloxane) (PDMS) or polyacrylate,[61] to collect volatiles from the headspace vapour. The chemical structure of the coating is selected such that the affinity of the target analytes towards the coating is optimised.[62] The collected sample can be subsequently analysed by GC or GC–MS. This technique is simple to use, requires no solvent, is relatively sensitive, and has been found to be more responsive to polar volatiles when compared with Tenax trapping. Hence, the application of SPME to the study of the Maillard reaction has been explored by many researchers and has become a popular sampling method in this area.[63-65]

6   In a Nutshell

The sampling technique selected is very dependent on exactly what the analysis is expected to achieve. For example, the use of an aggressive sampling method, such as the Soxhlet or Likens–Nickerson methods, may give a good representation of the volatile components of the whole foodstuff, but may not provide information about the aroma profile experienced by the consumer at the point of consumption. Also, the methodology used should be as mild as possible, so as to minimise the degradation of the target compounds and to prevent further reaction occurring. This again ensures that the volatiles collected are representative of the sample and not of the particular sampling technique used. When embarking on a new piece of research that requires extraction of Maillard reaction products from a food, it is worth considering the use of a variety of extraction methods and comparing the results. In this way, the researcher can maximise confidence that the results reflect the Maillard chemistry, not artefacts of any one individual extraction procedure.

7   Further Reading

T.H. Parliment, M.J. Morello and R.J. McGorrin, eds., *Thermally Generated Flavors: Maillard, Microwave and Extrusion Processes*, American Chemical Society, Washington DC, 1994.

A. Braithwaite and F.J. Smith, *Chromatographic Methods*, 5th ed., Blackie Academic and Professional, Glasgow, 1996.

8   References

1. Y. Motomiya, N. Oyama, H. Iwamoto, T. Uchimura and I. Maruyama, *Kidney Int.*, 1998, **54**, 1357.
2. M. Nakayama, G. Izumi, Y. Nemoto, K. Shibata, T. Hasegawa, M. Numata, K.M. Wang, Y. Kawaguchi and T. Hosoya, *Peritoneal Dialysis Int.*, 1999, **19**, 207.
3. A.G. Nerlich and E.D. Schleicher, *Atherosclerosis*, 1999, **144**, 41.
4. S. Sugiyama, T. Miyata, Y. Ueda, H. Tanaka, K. Maeda, S. Kawashima, C.V. De Strihou and K. Kurokawa, *J. Am. Soc. Nephrology*, 1998, **9**, 1681.
5. T. Miyata, N. Ishiguro, Y. Yasuda, T. Ito, M. Nangaku, H. Iwata and K. Kurokawa, *Biochem. Biophys. Res. Comm.*, 1998, **244**, 45.
6. C. Janzowski, V. Glaab, E. Samimi, J. Schlatter and G. Eisenbrand, *Food Chem. Toxicol.*, 2000, **38**, 801.
7. J.G. Wilkes, E.D. Conte, Y. Kim, M. Holcomb, J.B. Sutherland and D.W. Miller, *J. Chromatogr., A*, 2000, **880**, 3.
8. P.F.G. de Sa, J.M. Treubig, P.R. Brown and J.A. Dain, *Food Chem.*, 2001, **72**, 379.
9. Z. Guzel-Seydim, A.C. Seydim and A.K. Greene, *J. Dairy Sci.*, 2000, **83**, 275.
10. S. Drusch, V. Faist and H.F. Erbersdobler, *Food Chem.*, 1999, **65**, 547.
11. J.A. Gerrard, S.E. Fayle and K.H. Sutton, *J. Agric. Food Chem.*, 1999, **47**, 1183.
12. J.A. Gerrard, S.E. Fayle, P.K. Brown, K.H. Sutton, L. Simmons and I. Rasiah, *J. Food Sci.*, 2001, **66**, 782.
13. K. Hiramoto, K. Sekiguchi, K. Ayuha, R. Asoo, N. Moriya, T. Kato and K. Kikugawa, *Mut. Res.- Environ. Mutagenesis Related Subjects*, 1996, **359**, 119.
14. M.A. Glomb, D. Rosch and R.H. Nagaraj, *J. Agric. Food Chem.*, 2001, **49**, 366.
15. J.P. Yuan and F. Chen, *Food Chem.*, 1999, **64**, 423.
16. V. Fogliano, S.M. Monti, T. Musella, G. Randazzo and A. Ritieni, *Food Chem.*, 1999, **66**, 293.
17. J.M. Ames, A. Apriyantono and A. Arnoldi, *Food Chem.*, 1993, **46**, 121.
18. J.M. Ames, A. Arnoldi, L. Bates and M. Negroni, *J. Agric. Food Chem.*, 1997, **45**, 1256.
19. J.M. Ames, R.G. Bailey and J. Mann, *J. Agric. Food Chem.*, 1999, **47**, 438.
20. R.G. Bailey, J.M. Ames and J. Mann, *J. Agric. Food Chem.*, 2000, **48**, 6240.
21. P. Bersuder, M. Hole and G. Smith, *J. Am. Oil Chem. Soc.*, 1998, **75**, 181.
22. M.F. Wang, Y. Jin, J.G. Li and C.T. Ho, *J. Agric. Food Chem.*, 1999, **47**, 48.
23. J.M. de Bruijn and M. Bout, *Zuckerindustrie*, 2000, **125**, 604.
24. K. Hiramoto, X.H. Li, M. Makimoto, T. Kato and K. Kikugawa, *Mut. Res.-Genetic Toxicology and Environmental Mutagenesis*, 1998, **419**, 43.

25. B.R. Darcy, G.B. Rintoul, C.Y. Rowland and A.J. Blackman, *J. Agric. Food Chem.*, 1997, **45**, 1834.
26. R.E. Majors, *LC-GC*, 1999, **17**, S7.
27. R. Teranishi and S. Kint, *Flavor Sci. Discuss. Flavor Res. Workshop*, 1993, 137.
28. S.J. Risch and G.A. Reineccius, *ACS Symp. Ser.*, 1989, **409**, 42.
29. R.M. Smith, *Gas and Liquid Chromatography in Analytical Chemistry*, John Wiley, Chichester, 1988.
30. D.R. Knopp, *Handbook of Analytical Derivatisation Reactions*, John Wiley, New York, 1979.
31. K. Blau and J.M. Halket, *Handbook of Derivatives for Chromatography*, John Wiley, New York, 1993.
32. T.H. Parliment, *Food Sci. Technol. (NY)*, 1997, **79**, 1.
33. J.A. Pino and P. Borges. *Alimentaria (Madrid)*, 1999, **301**, 39.
34. T.-H. Yu, M.-S. Yang, L.-Y. Lin and C.-Y. Chang, *Food Sci. Agric. Chem.*, 1999, **1**, 129.
35. T.-H. Yu, Y.-N. Chen and L.-Y. Lin, *Contrib. Low- Non-Volatile Mater. Flavor Foods*, 1996, 227.
36. M. Negroni, A. D'Agostina and A. Arnoldi, *J. Agric. Food Chem.*, 2001, **49**, 439.
37. M. Cioroi, *Czech J. Food Sci.*, 2000, **18**, 103.
38. A. Chaintreau, *Flavour Fragrance J.*, 2001, **16**, 136.
39. S.T. Likens and G.B. Nickerson, *Proc. Am. Soc. Brew. Chem.*, 1964, 5.
40. V.A. Yaylayan, A. Keyhani and A. Wnorowski, *J. Agric. Food Chem.*, 2000, **48**, 636.
41. V.A. Yaylayan, *Am. Lab. (Shelton. Conn.)*, 1999, **31**, 30.
42. A. Wnorowski and V.A. Yaylayan, *J. Agric. Food Chem.*, 2000, **48**, 3549.
43. A. Wnorowski and V.A. Yaylayan, *J. Anal. Appl. Pyrolysis*, 1999, **48**, 77.
44. L.W. Kroh, W. Jalyschko and J. Haseler, *Starch*, 1996, **48**, 426.
45. W. Baltes, *J. Anal. Appl. Pyrolysis*, 1985, **8**, 533.
46. F.C.-Y. Wang, *J. Chromatogr., A*, 1999, **843**, 413.
47. T.P. Wampler, *J. Chromatogr., A*, 1999, **842**, 207.
48. C.Y. Chang, L.M. Seitz and E. Chambers, *Cereal Chem.*, 1995, **72**, 237.
49. A. Braithwaite and F.J. Smith, *Chromatographic Methods*, 5th ed., Blackie Academic and Professional, Glasgow, 1996.
50. R. Van Wijk, *J. Chromatogr. Sci.*, 1970, **8**, 418.
51. A. Zlatkis, H.A. Lichtenstein, A. Tishbee, W. Bertsch, F. Schumbo and H.M. Liebich, *J. Chromatogr. Sci.*, 1973, **11**, 299.
52. S. Muresan, M. Eillebrecht, T.C. de Rijk, H.G. de Jonge, T. Leguijt and H.H. Nijhuis, *Food Chem.*, 2000, **68**, 167.
53. M.E. Bailey, R.A. Gutheil, F.H. Hsieh, C.W. Cheng and K.O. Gerhardt in *Thermally Generated Flavours: Maillard, Microwave and Extrusion Processes*, eds. T.H. Parliment, M.J. Morello and R.J. McGorrin, American Chemical Society, Washington DC, 1994, 315.
54. C.-T. Ho and W.E. Riha III in *The Maillard Reaction in Foods and Medicine*, eds. J. O'Brien, H.E. Nursten, M.J.C. Crabbe and J.M. Ames, Royal Society of Chemistry, Cambridge, 1998, 187.
55. R. Villota and J.G. Hawkes in *Thermally Generated Flavors: Maillard, Microwave and Extrusion Processes*, eds. T.H. Parliment, M.J. Morello and R.J. McGorrin, American Chemical Society, Washington DC, 1994, 280.
56. J.W. Kinnison and R.S. Chapman, *Snack Food*, 1972, **61**, 40.
57. M. Nair, Z. Shi, M.V. Karwe, C.T. Ho and H. Daun in *Thermally Generated Flavours: Maillard, Microwave and Extrusion Processes*, eds. T.H. Parliment, M.J. Morello and R.J. McGorrin, American Chemical Society, Washington DC, 1994, 334.
58. C.L. Arthur, L.M. Killam, K.D. Buchholz, J. Pawliszyn and J.R. Berg, *Anal. Chem.*, 1992, **64**, 1960.
59. C.L. Arthur and J. Pawliszyn, *Anal. Chem.*, 1990, **62**, 2145.
60. A.D. James and G.A. Mills, *CAST, Chromatogr. Sep. Technol.*, 1999, **6**, 8.

61. H.H. Jelen, K. Wlazly, E. Wasowicz and E. Kaminski, *J. Agric. Food Chem.*, 1998, **46**, 1469.
62. Z.Y. Zhang and J. Pawliszyn, *Anal. Chem.*, 1993, **65**, 1843.
63. D.R. Cremer and K. Eichner, *J. Agric. Food Chem.*, 2000, **48**, 24.
64. M.J. Cantalejo, *J. Agric. Food Chem.*, 1997, **45**, 1853.
65. J.S. Elmore, M.A. Erbahadir and D.S. Mottram, *J. Agric. Food Chem.*, 1997, **45**, 2638.

CHAPTER 4

Gas Chromatography

1 Introduction

As was described in Chapter 1, the Maillard reaction is known to produce a complex array of flavour compounds that are responsible for both the taste and aroma of foods.[1,2] The characteristics of these compounds have long been an area of great interest since a thorough understanding of the mechanisms by which natural flavours are produced, lost or modified during food processing will assist in the creation of realistic food flavourings and enable food producers to minimise the development of off-flavours.

In Chapter 3, we discussed the various methods that might be employed to extract molecules from food, in order that the course and consequences of the Maillard reaction may be analysed. Having obtained a sample using a suitable extraction technique, it can be subjected to any of a wide range of separation techniques. One of the most popular techniques for studying Maillard reaction mixtures is gas chromatography (GC), especially capillary GC,[3-8] which is described in this chapter.

GC is a separation technique for volatile compounds that utilises very long chromatographic columns, usually from 10 to 100 metres in length, that contain a thin layer of a liquid stationary phase adhered to the inner wall of the column.[9]

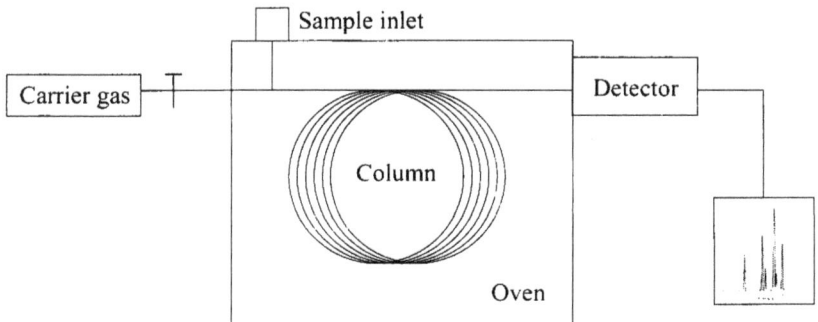

Figure 4.1 *Schematic of a GC instrument*

The reaction sample is separated into its individual components on the basis of both the volatility and the degree of interaction with the liquid stationary phase within the column. The sample is transported through the column by the flow of an inert carrier gas, for example nitrogen, hydrogen or helium, which eventually delivers the separated components to a detector, as shown in Figure 4.1.

2 The Separation

Having isolated a sample for analysis by GC, it is then necessary to make various decisions as to exactly how the separation of the sample components will be performed. The most important variables that must be considered include the injection method, the nature of the mobile and stationary phases and the oven temperature. Each of these variables will be addressed in turn.

The Injection

In order to ensure the success of the subsequent separation, it is desirable that the sample be introduced to the capillary column as a sharp band in the carrier gas stream. Samples can initially be in the solid, liquid or gas phase. Solid samples are generally pre-dissolved in a suitable solvent and injected as dilute solutions using a microsyringe, whereas those in the gas phase can be introduced under pressure or *via* a gas-tight syringe. The port is often sealed with a rubber septum, which reseals after the needle has been withdrawn. The injector port is usually held at a high temperature, so that the sample is immediately vaporised and carried into the column, where the sample components are separated.

For multiple samples, a programmable autosampler may be used to automatically remove and inject samples from a carousel. Autosamplers can be used in conjunction with samples that are extracted using the solid phase microextraction (SPME) or purge and trap sample isolation methods, discussed in Chapter 3, since a series of samples can be left in the carousel which are then automatically extracted and injected in series.

Capillary columns have a low sample capacity, therefore a number of injection techniques have been developed to avoid column overloading, with consequent band broadening and loss of resolution of the components being separated. The most common technique employs the split/splitless injector, which, as the name suggests, can be used either in the split mode or the splitless mode, as required for the specific analysis. During split injection, only a small proportion of the sample is injected into the column, usually just 1–10% of the sample, whilst the remainder of the sample is vented. This technique is ideal if only a limited number of components are present. For increased sensitivity, for instance when a sample contains trace compounds in low concentrations, a splitless technique is preferable. Under splitless conditions, most of the sample is injected. Hence, the splitless injector method is not suitable for highly concentrated samples, since the column would quickly become overloaded.

A modification of the splitless technique, which minimises band broadening, is cold trapping. This involves vapourising the sample as normal but, as the temperature of the column is held well below the boiling point of the sample, the sample is trapped on the first metre or so of the column. Once the injection is complete, the sample is re-vaporised by the progressive heating of the column.

A further technique, often employed for quantitative analyses, is cold on-column injection. Unlike the methods described above, where the sample is first volatilised and then passed onto the column, using this methodology the sample is placed directly into the cooled column in the liquid phase. As the oven, containing the GC column, is heated, so too is the injection port and the sample vaporises. This is a much gentler method of injection and, hence, is the preferred method of analysis for unstable compounds.

The Mobile Phase

The choice of carrier gas, that is, the mobile phase, is important both for the efficiency of the separation as well as for the performance of the detector, as discussed in Section 3. Carrier gases should be free from impurities, particularly air or water vapour, as these may introduce artefacts that will complicate the chromatogram. The most important choice, with regard to the resolution of the GC separation, however, is the choice of stationary phase.

The Stationary Phase

GC, or more correctly, gas–liquid chromatography (GLC), columns are typically comprised of a polyimide-coated fused-silica capillary, with an internal diameter of 25–75 µm, that has a thin coating of a liquid stationary phase on the internal wall of the column. Separations can also be performed using wide-bore columns, rather than capillary columns, or by gas–solid chromatography (GSC), using packed columns as in HPLC. As these columns have not normally been used for the analysis of Maillard reaction products, they will not be discussed here.

Literally hundreds of different liquid stationary phases of varying chemical compositions are commercially available.[10-12] These can, however, be grouped into a limited number of types: non-polar phases, such as those containing hydro-carbons or alkylsilicone polymers; polar phases, including polyether or substituted silicones; and specialised phases such as those used for the separation of amines or chiral molecules. The stationary phases most commonly used for GC analysis involve non-polar materials, such as methyl silicone.[13] Columns containing methyl silicone separate sample components by their boiling point – compounds with low boiling points are eluted first, while higher boiling compounds are retained by the column and elute last. Most columns, however, tend to separate on the basis of the chemical composition of the analyte, for example, they will separate on the basis of polarity or will have an affinity for a particular functional group, retaining any compounds that contain that group. The choice of column is, therefore, dependent on the chemical composition of the sample components to be separated.

Oven Temperature

A further parameter that must be considered when attempting to optimise a separation is the selection of an appropriate column temperature. The oven controls the temperature of the column and can be programmed to remain stable throughout the length of the separation. This is known as an isothermal separation and is usually only used for simple separations – a phrase not commonly associated with the separation of Maillard reaction products! If the sample components have varying volatilities, an isothermal separation will cause the later eluting components to suffer from band broadening, due to diffusion, resulting in broad peaks. One method of minimising this problem is to use a higher column temperature, hence speeding up the separation process and minimising diffusion effects. Although this may improve the peak shape of the later eluting components, it may prove detrimental to the resolution of those eluting earlier. For complex analyses, such as those resulting from the Maillard reaction, it is essential that a temperature gradient be utilised.[14 16] This involves the ramping of the temperature, from an initially low value, over a period of time. A temperature gradient may also include a number of isothermal periods (see Figure 4.2).

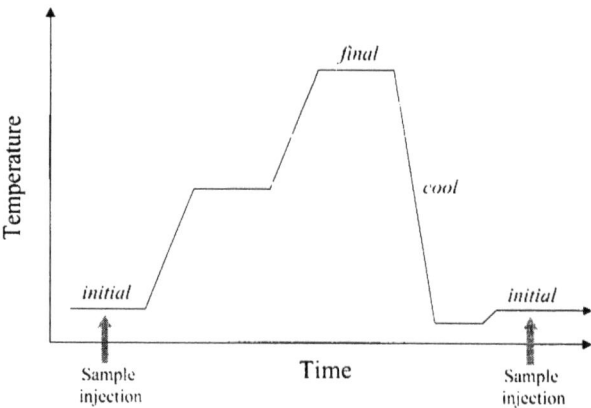

Figure 4.2   *Example of a GC temperature profile*

Gradually increasing the temperature will improve the peak shape of the later eluting components, reduce the retention time and the overall analysis time, without affecting the components that elute earlier. If a temperature programme is to be repeated, it is important that sufficient time be allowed for the oven to re-establish the starting temperature at the end of the cycle. Failure to do this will result in a higher starting temperature, altering the resolution of the earlier peaks.

3 Detection Methods

Having separated the components of a reaction mixture, the elution of the resolved analytes can be monitored by any of a wide range of detectors. The detectors most commonly associated with GC are discussed below.

Mass Spectrometry (MS)

The most powerful and widely used method of studying the volatile products of the Maillard reaction is through the combination of GC, for separation, with mass spectrometry (MS) as the detector.[17-20] As well as offering a means of detection, GC–MS systems provide information that can lead to the identification of unknown compounds. Detection techniques, commonly associated with GC, typically aid the identification of reaction components by their retention time only. As the majority of the products formed by Maillard chemistry are unknown, and authentic samples are not often available for comparative retention times, the use of GC–MS has provided an extremely useful means of furthering our knowledge in this field. A comprehensive discussion of MS, and its combination with GC, is given in Chapter 6 and will not be covered here. The following detection methods are, however, extremely useful if the retention time of a particular Maillard product of interest is known, and each has strengths in particular situations.

Sensory Evaluation – Gas Chromatography–Olfactometry (GCO)

Another common method of studying the volatile products of the Maillard reaction, specifically those contributing to the aroma profiles of foods, is GC–olfactometry (GCO).[21-23] GCO is used to locate the positions of odorants in a gas chromatogram and is achieved by the smelling of the humidified GC-effluent by trained personnel. In other words, human subjects are used as the detector. Trained assessors identify the odour-active areas of the chromatogram and record odour descriptions; for example, an odour could be described as cheesy, stale, floral or pungent, and the duration time of the smell is recorded. The importance of the various odour-active compounds to the overall aroma of the food can be estimated by a variety of methods, including aroma extract dilution analysis (AEDA)[24] or Charm analysis.[25] These techniques involve the analysis of a series of dilutions from the original aroma extract. The determination of the highest dilution at which the compound can still be detected provides an indication as to whether it is a potent odour compound[26] that should be prioritised for identification, by GC–MS for example. GCO is often used in conjunction with some other form of detection, such as those listed below.

Flame Ionisation Detector (FID)

A very popular GC detector in general is the flame ionisation detector (FID), shown in Figure 4.3. It can be used for the analysis of virtually all organic

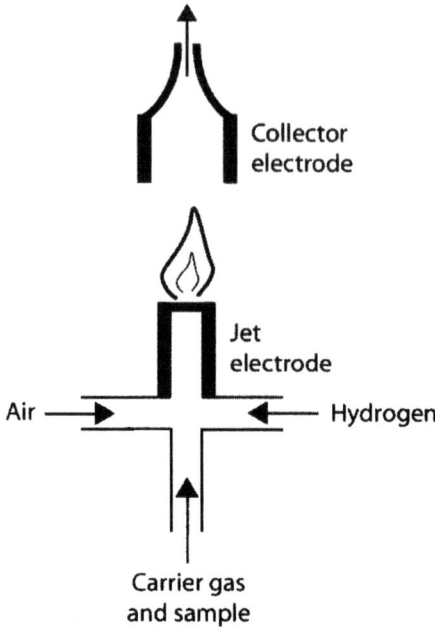

Figure 4.3   *Schematic of a FID*

compounds, due to its high sensitivity and reliability.[27] The FID generates ions through the combustion of organic compounds in a flame as they elute from the column, mixing the eluant gas with hydrogen and air, and burning it in a flame. A potential is applied across the flame at a collector electrode and any changes in the current, due to the elution of an analyte, are amplified and recorded. The main shortcomings of this detector are the need for three gases (the carrier gas, air and hydrogen) and, more importantly, the inability to identify an unknown compound. For the identification of unknown components of a mixture, GC–MS should be employed. Unlike GC–MS, however, the peak response of FID can be used to quantify the ratio of compounds, whereas the peak response of GC–MS is dependent on the degree of fragmentation of a compound, so is not as useful for estimating the relative concentration of sample components.

Other Detectors

A wide variety of GC detectors is available for the analysis of volatile compounds, but are not included here since they have not been used for the study of Maillard reaction systems. Others that have found occasional use include: the thermal conductivity detector (TCD), the electron capture detector (ECD), the flame photometric detector (FPD) and the nitrogen–phosphorus detector (NPD). Some of these are met in Case Study 4.1.

TCD measures changes in the thermal conductivity of the GC eluant. When the carrier gas flows through the detector cell, a thermal equilibrium is established.

When an analyte combines with the carrier gas and passes through the detector cell, the overall thermal conductivity of the eluant is altered and causes the temperature within the detector cell to change. This change in temperature is measured as a change in current that is subsequently amplified and results in the chromatographic signal. This detector is relatively insensitive, when compared with FID, since differences between the thermal conductivities of the carrier gas and the analyte are small.

ECD is a selective detector that has a high sensitivity toward halogenated compounds and other molecules that are able to behave as electron-acceptors, such as unsaturated or aromatic organic compounds and organometallic compounds. FPD is also a selective detector, but is highly specific for sulfur- and phosphorus-containing compounds.

Also known as the thermionic ionisation detector (TID) or the alkali flame ionisation detector (AFID), NPD is a modified version of the FID but, as the name suggests, is highly specific for compounds containing nitrogen and phosphorus. The selectivity of this detector can be altered so that it is specific either to phosphorus-containing compounds or to those containing both phosphorus and nitrogen.

CASE STUDY 4.1 – Flavour Components of Wine

Many researchers have studied the reactions of amino acids with carbonyl and dicarbonyl compounds. The majority of these studies, however, have concentrated on exploring the effect of food processing conditions, which traditionally involve high temperatures and high pH. The effect of lower temperatures and pH, such as those required during winemaking, have attracted much less attention. Pripis-Nicolau *et al.* have utilised GC in their study of the Maillard reaction under such conditions.[15]

A range of amino acids was reacted with a variety of carbonyl compounds in conditions that mimic the physicochemical conditions of wine. The volatile reaction products were isolated by solvent extraction, over a range of time periods, and were analysed by GC using five different detection methods – FID, NPD, FPD, MS and olfactometry. It was hoped that this combination of analytical methodologies would provide insight into the odorous compounds formed during bottling and storage.

When the sulfur-containing amino acids methionine and cysteine were reacted with each of the carbonyl compounds, potato and cabbage notes, with varying intensities, and rotten egg notes were described respectively. For methionine it was suggested that these olfactive descriptions were related to the formation of methanethiol from methional. Methanethiol can be oxidised to dimethyl disulfide, as shown in Figure 4.4. For cysteine, hydrogen sulfide and methanethiol were both thought to be produced.

The model reactions containing cysteine produced more complex and varied odours. These ranged from very intense popcorn and roasted odours to strong sulfurous and catty odours. As well as identifying a number of sulfur-

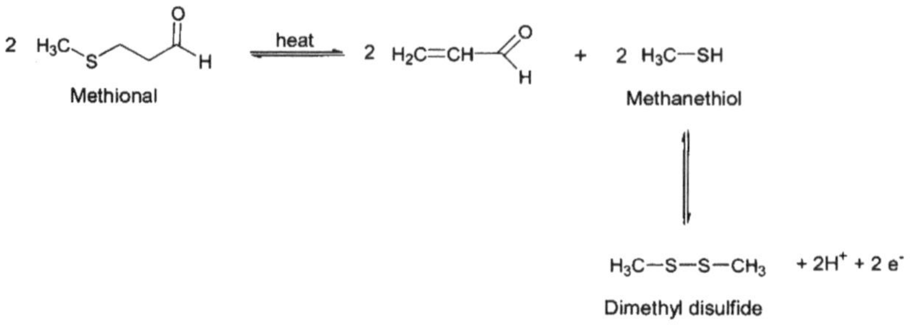

Figure 4.4   *Formation of methanethiol and dimethyl disulfide, adapted from Pripis-Nicolau et al.*[15]

containing products, many heterocycles, such as thiazoles, pyrazines and oxazoles, were identified. As these compounds all contain nitrogen, their presence could easily be confirmed with the use of the selective NPD and FPD detectors. Alkyloxazoles were identified in model reactions containing acetaldehyde. This compound is commonly associated with melon and very ripe kiwifruit notes. The authors proposed a possible pathway for the formation of alkyloxazoles, which is shown in Figure 4.5.

This work demonstrated the importance of Maillard chemistry under conditions encountered during the conservation and ageing of wines, such as low pH and low temperature.

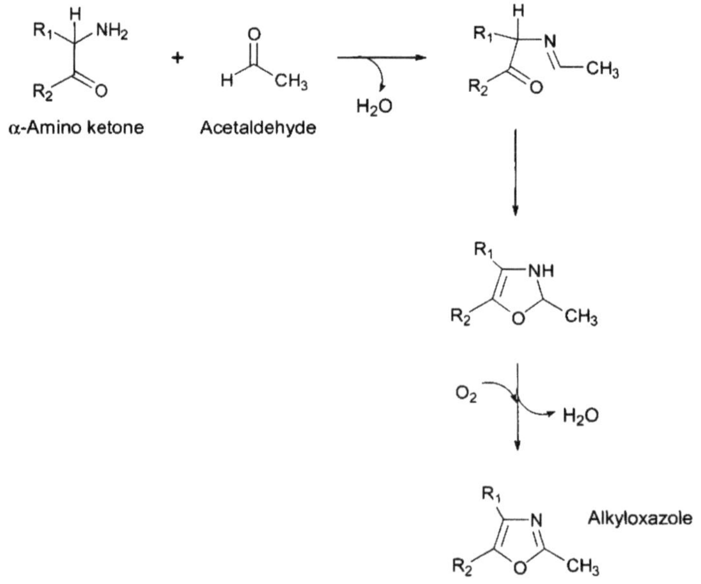

Figure 4.5   *Possible pathway for the formation of alkyloxazoles from acetaldehyde and α-amino ketones, adapted from Pripis-Nicolau et al.*[15]

4 Getting Started – in a Nutshell

As with most techniques, when developing a GC method a search of the literature should provide a useful starting point from which method optimisation can occur. The primary decisions that need to be made involve the choice of sampling technique, discussed in Chapter 3, injection method, carrier gas, column type and detector. The method used for the collection of volatiles, the injection technique, the selection of a column with an appropriate stationary phase and the oven temperature require the most thought, as it is these four parameters that can have the greatest impact on the selection of volatiles for analysis and the resolution of the separation.

The selection of the carrier gas can be limited by the choice of detector. For example, high sensitivity is best achieved using hydrogen or helium as the carrier gas when ECD is the method of detection. If this is not the case, oxygen-free nitrogen, or the slightly more expensive gas helium, are often the gases of choice since they are non-toxic, non-flammable and non-reactive. The flow rate of the gas affects the speed of the analysis and hence the resolution. Theoretically, the longer the sample is in the column, the better the separation. This is countered, however, by increased diffusion. The carrier gas flow rate will therefore require some manipulation so that the best separation is achieved in the minimum analysis time.

The most appropriate stationary phase is chosen after considering the characteristics of the sample – its polarity, the range of volatilities and the number of components. A useful starting point may be to select a stationary phase that has a polarity similar to that of the sample. This will result in the sample being well retained by the column, providing an adequate separation. If the polarity of the sample is not known, then analysis by both a polar and a non-polar column should provide some indication as to the polarity characteristics of the sample. Optimisation can then be achieved by repeated analyses with columns of slightly differing polarity. Up-to-date lists of commercially available columns, with their relative polarity, can be obtained from most suppliers. For most Maillard compounds, a non-polar column, such as 100% poly(dimethylsiloxane) (PDMS), should provide acceptable separations. Compounds which are poorly resolved by this column, namely alcohols, aldehydes and pyrazines, are likely to be better separated on a Carbowax-type column.

Once a column has been selected, it is then necessary to optimise the temperature of the oven. A gradient temperature programme should be used; this typically involves an initial period where the starting temperature is held constant, usually long enough for the solvent to elute. It is then raised at a selected rate, usually somewhere between 1 and 40 °C min^{-1}, and held at the final temperature for a short period of time. The maximum temperature at which a specific column can be used should be noted, as use above this temperature will result in degradation of the column. A typical starting programme may be a temperature gradient of 50–220 °C, raised at a rate of 4 °C min^{-1}, with a final hold of 10 min.

In many areas of separation science, GC has been somewhat eclipsed by more modern methodologies, such as HPLC (Chapter 5) and CE (Chapter 8). For many current studies in the Maillard field, especially those involving reaction products that are involatile, HPLC or CE may well prove more appropriate than GC. However, it is unlikely that GC will be entirely replaced, due to the enormous importance of volatile compounds in the flavour and aroma profiles of foods. GC is thus likely to remain a key 'ool for Maillard research for the foreseeable future.

5   Further Reading

D. Rood, *A Practical Guide to the Care, Maintenance and Troubleshooting of Capillary Gas Chromatographic Systems*, 3rd ed., Wiley, New York, 1999.

H.M. McNair and J.M. Miller, *Basic Gas Chromatography: Techniques in Analytical Chemistry*, Wiley, New York, 1998.

L. Matter (ed.), *Food and Environmental Analysis by Capillary Gas Chromatography: Hints for Practical Use*, Huethig, Heidelberg, 1997.

R. Wittkowski and R. Matissek (eds.), *Capillary Gas Chromatography in Food Control and Research*, Technomic, Lancaster, 1993.

6   References

1. H. Weenen, J. Kerler and J.G.M. van der Ven in *Flavours and Fragrances*, ed. K.A.D. Swift, Royal Society of Chemistry, Cambridge, 1997, 153.
2. H.E. Nursten in *Development of Food Flavours*, eds. G.G. Birch and M.G. Lindley, Elsevier Applied Science, London, 1986, 173.
3. M. Pischetsrieder, B. Huber and W. Seidel, *Lebensmittelchemie*, 1999, **53**, 144.
4. T. Hofmann and P. Schieberle, *J. Agric. Food Chem.*, 1995, **43**, 2187.
5. A. Keyhani and V.A. Yaylayan, *J. Agric. Food Chem.*, 1996, **44**, 2511.
6. D.S. Mottram and F.B. Whitfield, *J. Agric. Food Chem.*, 1995, **43**, 984.
7. T.H. Parliment, *ACS Symp. Ser.*, 1998, **705**, 8.
8. R. Tressl, G.T. Wondrak, R.P. Kruger and D. Rewicki, *J. Agric. Food Chem.*, 1998, **46**, 104.
9. A. Braithwaite and F.J. Smith, *Chromatographic Methods*, 5th ed., Blackie Academic and Professional, Glasgow, 1996.
10. D.O. Popescu, P. Ionita, N. Zarna, I. Covaci, A. Stoica, A. Zarna, D. Nourescu, F. Spafiu, M.T. Caproiu, C. Luca, F. Badea, T. Constantinescu and A.T. Balaban, *Roum. Chem. Q. Rev.*, 1998, **6**, 271.
11. N. Fukunaga, *Kuromatogurafi*, 1995, **16**, 104.
12. G. Guiochon, *Adv. Chromatogr.*, 1969, **8**, 179.
13. M. McMaster and C. McMaster, *GC/MS – A Practical Users Guide*, Wiley–VCH, New York, 1998.
14. W.L.P. Bredie and D.S. Mottram, *Food Chem.*, 1995, **55**, 109.
15. L. Pripis-Nicolau, G. De Revel, A. Bertrand and A. Maujean, *J. Agric. Food Chem.*, 2000, **48**, 3761.
16. V.M. Hill, N.S. Isaacs, D.A. Ledward and J.M. Ames, *J. Agric. Food Chem.*, 1999, **47**, 3675.
17. E. Valero, M. Villamiel, B. Miralles, J. Sanz and I. Martinez-Castro, *Food Chem.*, 2001, **72**, 51.
18. Y. Zheng, S. Brown, W.O. Ledig, C. Mussinan and C.-T. Ho, *J. Agric. Food Chem.*, 1997, **45**, 894.
19. M.W. Samsudin, R.T. Sun and I.M. Said, *J. Agric. Food Chem.*, 1996, **44**, 247.

20. J.K. Parker, G.M.E. Hassell, D.S. Mottram and R.C.E. Guy, *J. Agric. Food Chem.*, 2000, **48**, 3497.
21. M.D. Aaslyng, J.S. Elmore and D.S. Mottram, *J. Agric. Food Chem.*, 1998, **46**, 5225.
22. W.L.P. Bredie, D.S. Mottram and R.C.E. Guy, *J. Agric. Food Chem.*, 1998, **46**, 1479.
23. Y. Wang and S.J. Kays, *J. Am. Soc. Hort. Sci.*, 2000, **125**, 638.
24. W. Grosch, *Trends Food Sci. Technol.*, 1993, **4**, 68.
25. T.E. Acree, J. Barnard and D.G. Cunningham, *Food Chem.*, 1984, **14**, 273.
26. T. Hofmann and P. Schieberle, *J. Agric. Food Chem.*, 1997, **45**, 898.
27. R.M. Smith, *Gas and Liquid Chromatography in Analytical Chemistry*, Wiley, Chichester, 1988.

CHAPTER 5

Liquid Chromatography

1 Introduction

In Chapter 4, we discussed the various methods that might be employed to separate Maillard reaction products *via* gas chromatography (GC), in which the molecules were partitioned between a carrier gas (the mobile phase) and a liquid (the stationary phase). In this chapter, we will examine a similar technique – liquid chromatography. The two methods are closely related, but in the case of liquid chromatography, molecules are separated by partitioning between a solvent (the mobile liquid phase) and a solid support (the stationary phase), usually in the form of a packed column. Many of the separation principles underlying GC are also relevant here, but the details of the separation protocols are rather different.

High pressure (or high performance) liquid chromatography (HPLC), in one form or another, is perhaps the most common general method applied to the separation of molecules from food mixtures and in many other scientific endeavours. This is reflected in the increasing number of excellent texts that are available, describing the principles and practice of liquid chromatography (see Further Reading). Not surprisingly then, there are very many examples of the use of HPLC in the Maillard arena and, as a research tool, HPLC is becoming an increasingly important part of many research programmes.[1-12]

The details of particular procedures used to separate molecules by liquid chromatography are many and various, but the basic principles are the same throughout. HPLC has its roots in the more traditional forms of liquid chromatography – thin layer chromatography (TLC) and low pressure column chromatography. These are each discussed below, to illustrate some basic principles of separation, before considering how these principles are applied to the analysis of the Maillard reaction by HPLC in food science.

2 Separation Basics

Thin Layer Chromatography

Liquid chromatography has long been carried out on a matrix that is supported as a thin layer, in TLC. Thin layer chromatography is becoming increasingly outmoded, although it still finds applications,[13] including some within the Maillard field.[14–19] Case Study 5.1 describes a recent application of TLC to identify Maillard reaction products, in which it is used to complement a range of more sophisticated methodologies.

CASE STUDY 5.1 – Identification of Aminophospholipid-linked Maillard Compounds in Egg Yolk

Utzmann and Lederer[7] have recently published a study in which they compare the results of model studies with the identification of Maillard products in a processed food – spray-dried egg yolk. They used a raft of techniques, including liquid chromatography with mass spectrometric detection (LC–MS, see Section 4), to identify some hitherto unidentified Maillard reaction products of aminophospholipids. Many Maillard reaction products were found in egg yolk, including phosphatidylethanolamine-linked glucosylamines, Amadori products and 5-hydroxymethylpyrrole-2-carbaldehydes. Figure 5.1 shows a TLC separation of three egg yolk preparations, clearly identifying a glycated phosphatidylethanolamine derivative with a unique R_f, later identified by LC–MS. Hence, the existence of this compound in a food sample can be clearly demonstrated by a simple TLC analysis.

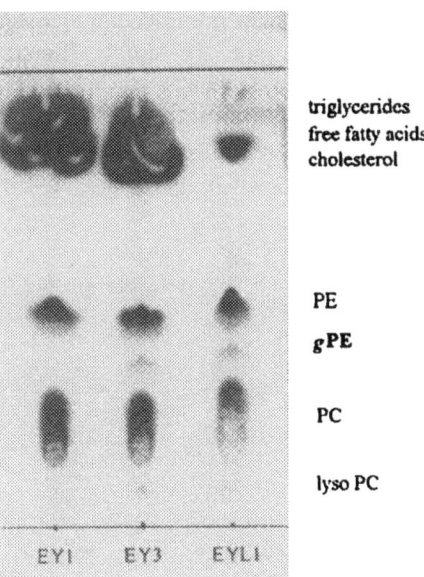

triglycerides
free fatty acids
cholesterol

PE
*g*PE

PC

lyso PC

EY1   EY3   EYL1

Figure 5.1 *TLC separation of egg yolk preparations showing a unique spot which corresponds to glycated phosphatidylethanolamine (gPE). The three lanes (EY1, EY3, EYL1) represent different spray-dried egg yolk samples. Phosphatidylethanolamine (PE) and phosphatidylcholine are also detected by this method.*
(Reproduced with permission[7])

Such glycated derivatives of phosphatidylethanolamine are predicted to influence the emulsifying properties of egg yolk and the resistance of egg products to oxidation.

Liquid Chromatography

Most modern liquid chromatography is carried out by column HPLC. However, before discussing some of the finer points of this enormously important technique, it is worth pausing to reflect on some basic principles that govern much of the theory behind the most commonly practised of these highly sophisticated protocols that we see detailed in the literature. Liquid chromatography of molecules allows them to be separated by differences in their physicochemical properties. Most of these derive from two simple molecular properties – size and charge – each of which will be briefly described for the simplest possible low-pressure column.

Separation by Size

Size exclusion chromatography, also known as gel filtration or gel permeation chromatography, relies on differences in molecular size and, to some extent, shape to afford separation. The principles are simply illustrated for a low-pressure column in Figure 5.2. Fine porous beads are packed into a column, and the molecules to be separated are poured onto the column in a small volume of an appropriate solvent. Large molecules are excluded from the beads, and elute from the column first. Small molecules enter the beads, and their progress is retarded.

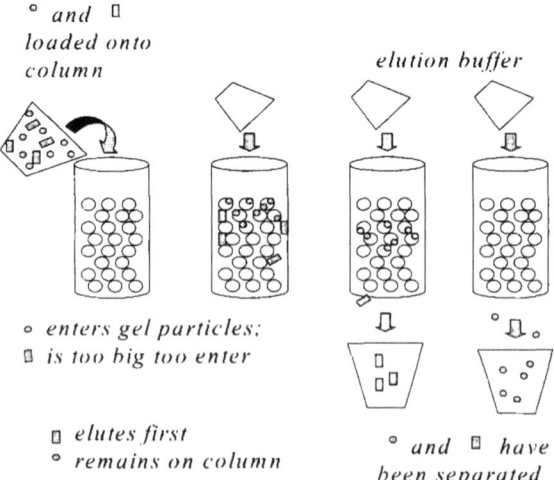

Figure 5.2   *Size exclusion chromatography – schematic*

Thus they elute from the column later on in the experiment, and are separated from the larger molecules.

Size exclusion chromatography is particularly useful for the separation of a range of molecules of widely different sizes. It ought, therefore, to find use in the separation of crude Maillard reaction products, where some molecules remain in the low molecular weight range, while others have polymerised (*e.g.* to melanoidins). There are examples in the literature where Maillard reaction products have been separated according to size,[1] by using dialysis[20] or ultrafiltration.[21] Despite this, size exclusion chromatography has not found wide application in the Maillard field.

Separation by Charge

Many Maillard reaction products have very similar sizes, and must therefore be separated on the basis of another property, often related to their charge. The charge of a molecule, and the distribution of charge within the structure of a molecule (its polarity), affects its affinity towards a charged, or polar, support. Charge distribution also affects the solubility of a molecule in solvents of various polarities, ionic strengths or pH. Thus, if a mixture of molecules is loaded onto a charged support, careful manipulation of the exact nature of the eluting solvent can effect separation of the mixture. This is illustrated schematically in Figure 5.3 for a simple mixture of two compounds. The column depicted contains sulfonic acid groups, which exist as anions at neutral pH. The column is, therefore, a cation exchanger. In the figure, cations A^+ and C^+ are separated in a series of manual steps. In the first instance, A^+ cations bind to the column, displacing H^+ ions, which elute with C^+ ions. Once all the C^+ ions have eluted, the A^+ ions are removed from the column by increasing the concentration of H^+ – thus the two cations are separated.

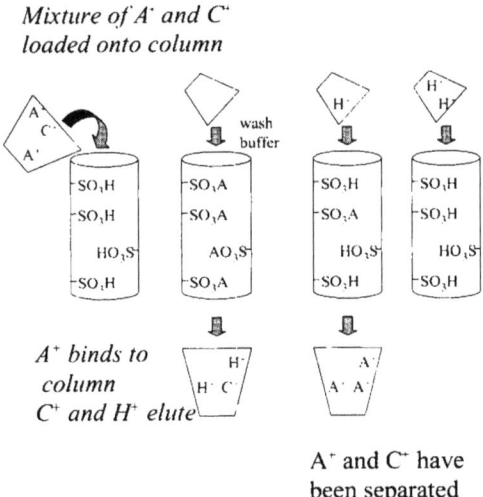

Figure 5.3 *Ion exchange chromatography – schematic*

Low pressure chromatography based on ion exchange has found application in the Maillard field, for example in early work elucidating the reaction products in glucose–lysine model systems.[22] A more recent example of low pressure chromatography successfully applied to the separation of Maillard reaction products is discussed in Case Study 5.2.

CASE STUDY 5.2 – Identification of Novel Maillard Products from the Reaction of Tryptophan and Xylose

Wang *et al.*[23] have recently published a study of the model reaction between tryptophan and xylose. The two compounds were refluxed in aqueous solution for 20 hours and the resulting non-volatile products extracted into ethyl

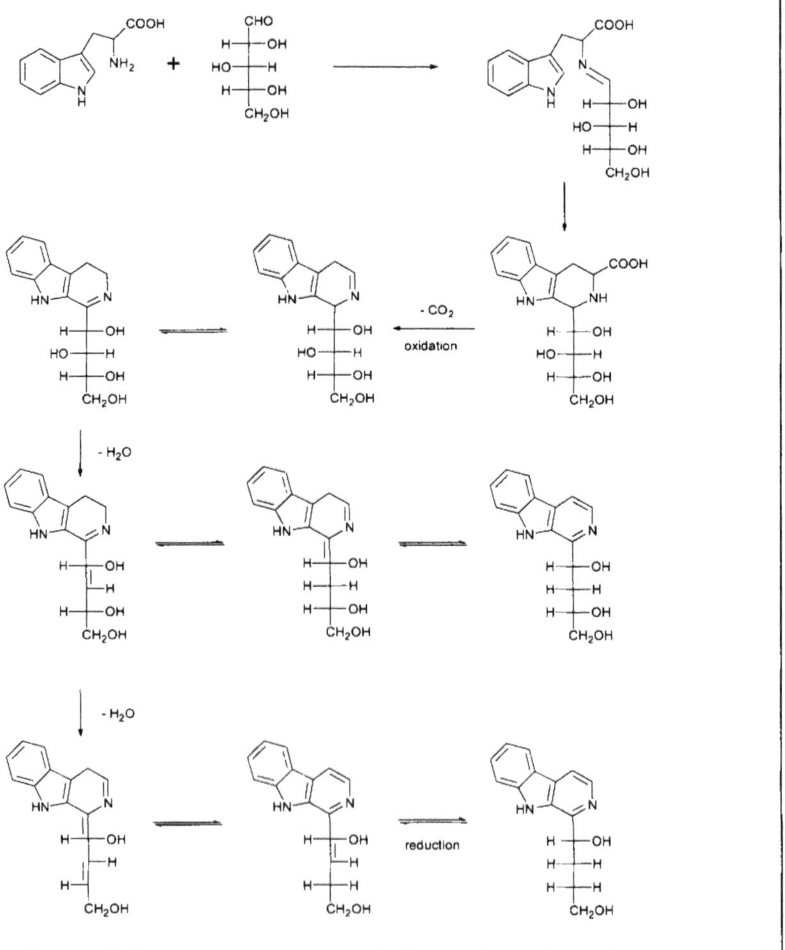

Figure 5.4 *Proposed formation of two novel β-carbolines from the reaction of tryptophan and xylose. Modified from Wang et al.[23]*

acetate. Separation of the Maillard reaction products was afforded by simple low-pressure chromatography on flash silica gel. Two novel compounds were identified by mass spectrometry and nuclear magnetic resonance – both of which had a β-carboline structure. The proposed pathway for formation of these structures is depicted in Figure 5.4.

Uncharged molecules can also be separated according to their affinity for a solid support. This principle is commonly exploited in reverse-phase chromatography, in which the column matrix is hydrophobic, rather than charged (or polar), and binds non-polar molecules. In this case, careful adjustment of the chosen organic solvent will effect separation of the molecules. Reversed-phase chromatographic separation of Maillard products is one of the most commonly used techniques and will be illustrated in the following sections.

3 The Art of Chromatography

To the novice, and sometimes even to the expert, liquid chromatography can appear as more of an art than a science, with a wide range of parameters to optimise for each new separation challenge. There is certainly no substitute for experience when it comes to setting starting conditions for a new separation procedure for any given collection of molecules. However, a systematic consideration of each variable assists any researcher in the development of a separation protocol. By way of a gentle introduction to this daunting field, each of these variables is considered in turn in the following discussion.

Low Pressure or High Pressure?

As briefly described above, low pressure column chromatography involves the packing of relatively large beads into a simple column, and using gravity to elute the solvent. It is simple, but relatively slow, and therefore only useful for those that have simple separations to effect, on a relatively infrequent basis. Thus, low pressure chromatography is of limited use to the typical Maillard researcher. For many years, low pressure chromatography was the only type of liquid chromatography available to separation scientists. Pressurising columns containing a matrix designed to run at low pressure was often counterproductive – it generally lead to the matrix compressing to such an extent that further passage of solvent was no longer possible. This problem was compounded for small bead sizes. Thus, although it was known that separation of molecules is improved by reducing the size of the beads in the column, bead sizes beyond a certain minimum were impractical due to the flow problems resulting from their compression, even at atmospheric pressure.[24]

In the 1970s, a breakthrough in packing technologies facilitated the development of high-efficiency columns that could run at high pressure. Furthermore, the small silica particles (10 μM) used for the packing of these columns were found to be amenable to surface modification procedures that specifically modified the binding characteristics of the surface. The contemporaneous development of these new technologies paved the way for the introduction of a technique that is now the dominant separation tool in many industries – HPLC.[24] Suddenly, the scientific community possessed sufficient separation power to tackle a broad range of problems in a reasonable analysis time. Unlike the already popular gas chromatography (GC, see Chapter 4), HPLC could deal with a broad range of samples that required little sample preparation. In principle at least, anything that can be dissolved is amenable to HPLC analysis.[24] High pressure chromatography has, therefore, almost completely superseded its low pressure counterpart, especially in situations where complicated mixtures of products must be separated, as is commonly encountered in the Maillard field.

Which Column?

There is a frightening array of HPLC columns currently available on the market. Having decided to embark upon an HPLC separation, one could spend several days reading through the available literature describing the apparently endless features of each particular column. Each type of column is tailored to meet the needs of certain types of separation challenge – usually of known molecules. Choosing a column to separate as yet unidentified compounds involves as much luck as judgement, and researchers often have to try many different systems before finding one optimal for their particular situation. As a starting point, the choice is somewhat simplified if the columns are categorised according to the principles on which they separate the molecules. These are related to the basic principles that we have already encountered in Section 2 – charge and size. Other factors that must be considered include column dimension. Each of these considerations is discussed briefly below.

Separation by Charge

Traditional low pressure liquid chromatography often employed a column containing a polar adsorbent (often silica, as in Case Study 5.2, or alumina) and a non-polar solvent (such as petroleum ether or dichloromethane). Molecules were separated according to their relative affinity for the column matrix and the solvent, commonly described as the way in which they 'partition between the polar stationary phase and the non-polar mobile phase'.[24] With a polar adsorbent, very non-polar molecules elute first and very polar molecules elute last. This type of chromatography is generally referred to as 'normal-phase chromatography' or 'adsorption chromatography', and can be used in high pressure columns. Ion exchange chromatography, as discussed before (Section 2) may also be carried out by HPLC. Here the column comprises charged groups that strongly interact with molecules of opposite charge present in the solution. Elution is achieved by changing the ionic strength or the pH of the eluting buffer, which may alter the charge of either the column, or the molecules to be separated.

Normal phase and ion exchange chromatographies are not the most common types of HPLC, although there are examples of their successful use. An example of a Maillard study that uses one type of normal-phase HPLC – in this case ion exchange HPLC (IE-HPLC) – and compares it to other separation techniques, is given in Case Study 5.3. Further examples include methods that have been developed to determine aromatic Amadori products.[25]

Normal-phase chromatography has been somewhat eclipsed by its *alter ego*, 'reverse-phase chromatography', in which the column support (stationary phase) is non-polar and the solvent is relatively polar. In reverse-phase chromatography (RP-HPLC), the polar molecules elute first, and the non-polar molecules are retained on the column for longer times. This type of chromatography is often, in practice, easier to use than normal-phase separations. This is attributed to the fortuitously rapid equilibration of molecules between the stationary and mobile phases, and the fact that retention times are often more reproducible. RP-HPLC is currently the most popular form of liquid chromatography, with about 90% of published separations of low molecular weight molecules employing this method.[24] Separation of Maillard reaction products is no exception, and there are innumerable studies in Maillard chemistry that employ RP-HPLC at some point.

RP-HPLC owes its popularity to the fact that the separation principle is well understood, and that good results can be obtained for a broad range of compounds with few technical complications.[24] It is ideal for the separation of as yet unknown compounds, and is consequently the method of choice for many Maillard studies where the precise nature of the products formed and their relative distributions is unlikely to be known.

Molecules are retained on a reversed-phase column by interaction with a hydrophobic surface. This surface is most commonly composed of a silica-based matrix with attached hydrocarbon chains – typically 8 or 18 carbon atoms long – hence common columns used to effect separation by RP-HPLC are often referred to as C8 and C18 columns, respectively. There are many other types of RP-HPLC column available, but those based on C8 and C18 matrices make good starting points for most separations, and have been used in many Maillard studies.[26-30]

CASE STUDY 5.3 – Analysis of a Model Extrusion-cooked Cereal Product

Ames *et al.*[31] undertook a comprehensive comparison of various separation techniques in their analysis of a model food system prepared by extrusion of a starch, glucose and lysine mixture. The resulting compounds were extracted with methanol and separated by isoelectric focusing (IEF, see Chapter 7), capillary zone electrophoresis (CZE, see Chapter 8), RP-HPLC and IE-HPLC. Selected reaction products were identified and characterised. CZE and RP-HPLC were found to establish differences in the profiles of molecules from mixtures that had been treated under different conditions – these two techniques also effected the best separation in the quickest timeframe.

Separation of three key components of the extruded mixture by RP-HPLC, using a type of C18 column, is shown in Figure 5.5.

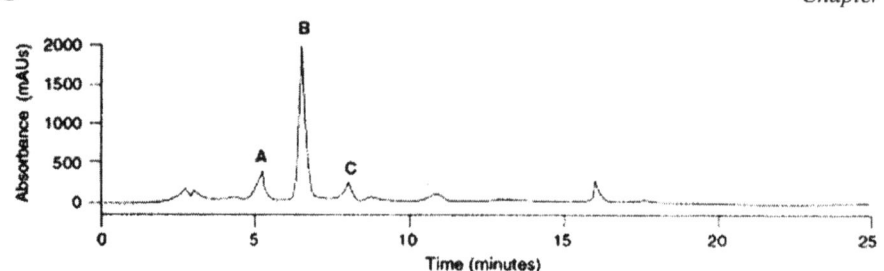

Figure 5.5 *Separation of three components of a model extrusion-cooked cereal product. Peak A was partially characterised, peak B was conclusively identified as 4-hydroxy-2-(hydroxymethyl)-5-methyl-3(2H)-furanone, peak C was 5-(hydroxymethyl)furfural*
(Reproduced with permission[31])

IE-HPLC was able to demonstrate clear differences between samples extruded under different conditions, but the analysis time was prohibitively long. Difficulties were also encountered due to the need to include buffers in the elution solvent.

Some molecules are unstable when exposed to the conditions required for RP-HPLC and more gentle conditions are required to separate mixtures containing these compounds. Perhaps the best example of this is in the separation of proteins, which may be denatured by reverse-phase chromatographic techniques. In instances such as these, the hydrophobicity of the solvent may be reduced and the molecules eluted with water or aqueous buffer. This is known as 'hydrophobic interaction chromatography', and is beginning to find use in the analysis of food proteins.[32-35] Hydrophobic interaction chromatography offers great potential for analysis in the Maillard field, where there is a need to expand the repertoire of separation techniques to include a greater focus on high molecular weight molecules. Currently this problem is circumvented by the digestion of large molecules into their components, followed by analysis of the small molecular weight products. Whilst this approach can be powerful and yield much information (see Case Study 5.4), it is plagued by the inevitable possibility that the breakdown of macromolecules prior to analysis may produce artefacts (see Chapter 3).

CASE STUDY 5.4 – Detecting Protein-bound Lysylpyrrolaldehyde in Dried Pasta

Pasta drying represents a relatively simple form of whole food processing in which to study Maillard browning, and was the subject of a detailed study by Resmini and Pellegrino.[36] It had previously been shown that Maillard chemistry takes place both between water-soluble components of the pasta and between water-soluble components and the protein matrix. One of the reaction products was found to be protein-bound lysylpyrrolaldehyde, and this was used as a marker to measure the extent of the Maillard reaction during pasta drying.

In both model systems and actual pasta, enzymatic hydrolysis was used to digest the protein and the resulting hydrolysate was separated by RP-HPLC. Figure 5.6 shows the results for the Maillard reaction products extracted from pasta – the lysylpyrrolaldehyde peak can be seen clearly.

Use of radioactively labelled glucose was also used in order to trace the origins of this reaction product. Such tracer techniques will be met in Chapter 9.

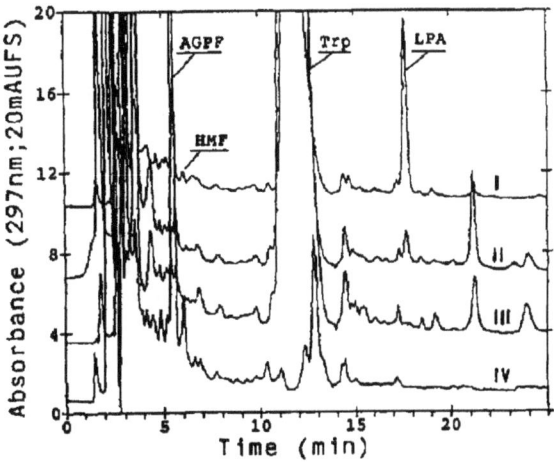

Figure 5.6   *HPLC elution profile of pasta processed under different heating conditions. Enzymatic hydrolysates of spaghetti dried with high temperature (line I); medium temperature (line II); and low temperature (line III). Water extract of experimental pasta containing protein-bound LPA (line IV). APGF = acetylglucopyranosylofuran, HMF = hydroxymethylfurfural, Trp = tryptophan, LPA = lysylpyrrolaldehyde.*
(Reproduced with permission[36])

Separation by Size

Size-exclusion chromatography, described in Section 2, may also be carried out under high pressure. This technique is most often applied to the separation of industrial polymers and biopolymers, but by using a range of pore sizes may be used for smaller molecules. This technique, although not yet widely investigated in the Maillard context,[37] shows great promise for the separation of molecules of different sizes from the Maillard reaction.

Column Dimensions

Most HPLC analyses are currently carried out in a column that has support material packed within a stainless steel tube, which varies in length from 10 to 25 cm and has a bore of 3.9–8.0 mm. Many workers have demonstrated that similar separations can be achieved on much narrower columns – with a bore of less than a millimetre. Such columns are known as microbore columns.[38] (Very narrow bore columns – capillaries – are considered in Chapter 8.)

Microbore columns have the advantage of reducing both the amount of sample required and the amount of solvent needed to elute that sample. Disadvantages include the lack of reliability in packing such narrow columns and the current lack of appropriate low-volume equipment with which to run the columns and detect the molecules in the eluant. However, these technical obstacles are increasingly being overcome, and many predict that HPLC methods of the future will rely heavily on microbore technology.[38] Microbore methods have already found application in the analysis of the Maillard reaction of milk proteins.[39,40]

Which Solvent?

When a new HPLC protocol is developed, a lot of time tends to be spent deliberating over the column that will be used to effect the separation. This is understandable, since the column has been shown to play the dominant role in the overall separation process[41] and represents a significant financial investment. However, the choice of solvent should not be overlooked, since the overall separation quality, especially peak shape, group specificity and many other operating parameters will also be critically dependent on the liquid phase.[41]

A thorough analysis of the effect of solvent on HPLC separation is beyond the scope of this book. However, in any individual study, researchers can derive great benefit from the optimisation of their solvent system for maximum separation and ease of characterisation. A few parameters that should be considered are: interaction of the solvent with the detection method (especially for UV-visible detection, see below); viscosity, which will affect the pressure of the system; volatility, which may affect the propensity of bubbles to form during the separation and lead to artefactual peaks; instability and reactivity of the solvent with the column and reaction products in question; and miscibility with co-solvents if a gradient elution is to be undertaken.[41]

Common solvents employed for HPLC separation of Maillard reaction products include water, alcohols, acetonitrile and chlorinated hydrocarbons. If a solvent is used 100% pure, the term 'isocratic elution' is generally used. In many studies, more effective separation can be achieved by using two or more solvents in different and varying ratios. This is known as 'gradient elution'. Most HPLC systems are set up in such a way that the solvent composition can be pre-programmed to alter in a defined manner, greatly improving the possible resolution that can be achieved for each mixture of compounds. Another solvent variable that can be varied is pH. Trifluoroacetic acid (TFA) is a common addition to HPLC solvents and can have excellent results (although it is worth bearing in mind that addition of acid may change the nature of the molecules being separated – a fact that is particularly relevant in the Maillard field).[41]

4   Detection Methods

Whatever the experimental system chosen for an HPLC separation, careful thought must be given to the method that will be used to detect the molecules as they elute from the column. There are many factors to bear in mind, not least of which is the availability of the detectors. A brief discussion of commonly used detectors is given below.

UV-visible Detection

By far the most common method of detection for HPLC separation is by exploiting the UV-visible absorption of the eluting compounds. This detection method has been used for the vast majority of Maillard studies that employ HPLC separation.[10,25,39,42-55] It is worth remembering that the specific wavelength chosen for detection will influence the number of products detected – not all molecules absorb at the same wavelength. A complete analysis is possible if photodiode array detection is available, which records the complete UV-visible spectrum for all eluting compounds. This has been employed in several Maillard studies.[48,55,56]

Refractive Index Detection

The refractive index of a solvent changes if the concentration of solutes within it is altered. Thus, any molecule can potentially be detected using a refractive index detector. This can be an advantage for some applications, for example in the detection of sugars, which cannot readily be detected by UV-visible detectors since they do not have very distinctive spectrophotometric absorbances. It can also be advantageous if every component in a sample needs to be located. Refractive index detection of molecules from HPLC separations of many food systems has been reported.[57-64] However, given the complexity of Maillard reaction product mixtures, and the ready detection of many components by UV-visible absorption, refractive index detection has not been employed widely in the Maillard field.

Fluorescence

As mentioned in Chapter 2, fluorescence is a molecular property that is increasingly associated with Maillard reaction products.[65-69] This opens up the possibility of employing fluorescence detection for HPLC analysis of Maillard reaction products, if an appropriate detector is available.[55,70]

Mass Spectrometry

As was the case with GC (Chapter 4), the coupling of liquid chromatography with mass spectrometry provides a powerful analytical combination that not only detects molecules, but simultaneously provides structural information to aid in the characterisation of the molecule. Such information is otherwise much harder to obtain, and often requires separation of the compound on a preparative scale, before chemical analysis (see Chapter 9). Liquid chromatography–mass spectrometry (LC–MS) is an increasingly popular technique amongst analytical food chemists, with huge potential in the Maillard field.[1,71-73] Mass spectrometric detection will be considered in Chapter 6.

5   Getting Started – in a Nutshell

Most research laboratories have ready access to HPLC equipment and a variety of columns. Modifying a literature method is often readily achievable with a basic knowledge of the separation principles involved in each type of column. For many purposes in the Maillard field, a good starting point would be an RP-HPLC separation using a C8 or C18 column, with UV-visible detection. A variety of different solvent systems should be surveyed in order to optimise separation of the compounds of interest.

Liquid chromatography is an essential tool for all Maillard researchers, and progress in the elucidation and characterisation of Maillard products from food will continue to be made as HPLC separation protocols and detection methods become ever more sophisticated. There is also great potential in employing more than one detection method simultaneously, in order to begin to cross-correlate various literature studies, and piece together more and more information about individual components of specific reaction systems.

6   Further Reading

U.D. Neue, *HPLC Columns: Theory, Technology and Practice*, John Wiley, New York, 1997.

J.E. Kruger and J.A. Bietz (eds.), *HPLC of Cereal and Legume Proteins*, American Association of Cereal Chemists, Inc., St Paul, Minnesota, 1994.

R.W.A. Oliver, *HPLC of Macromolecules: A Practical Approach*, 2nd ed., IRL Press, Oxford, 1998.

P.C. Sadek, *The HPLC Solvent Guide*, John Wiley, New York, 1996.

7   References

1. S.M. Monti, A. Ritieni, G. Graziani, G. Randazzo, L. Mannina, A.L. Segre and V. Fogliano, *J. Agric. Food Chem.*, 1999, **47**, 1506.
2. R.H. Nagaraj, M. Porterootin and V.M. Monnier, *Arch. Biochem. Biophys.*, 1996, **325**, 152.
3. P. Bersuder, M. Hole and G. Smith, *J. Am. Oil Chem. Soc.*, 1998, **75**, 181.

4. M.P. Ennis and D.M. Mulvihill, *Int. J. Dairy Technol.*, 1999, **52**, 156.
5. M.O. Lederer and R.G. Klaiber, *Bioorg. Med. Chem.*, 1999, **7**, 2499.
6. C.M. Utzmann and M.O. Lederer, *Carbohydr. Res.*, 2000, **325**, 157.
7. C.M. Utzmann and M.O. Lederer, *J. Agric. Food Chem.*, 2000, **48**, 1000.
8. P.A. Harmon, W. Yin, W.E. Bowen, R.J. Tyrrell and R.A. Reed, *J. Pharm. Sci.*, 2000, **89**, 920.
9. J.M. Ames, *Food Chem.*, 1998, **62**, 431.
10. T. Hofmann, *Carbohydr. Res.*, 1998, **313**, 203.
11. M.A. Glomb, D. Rosch and R.H. Nagaraj, *J. Agric. Food Chem.*, 2001, **49**, 366.
12. Q. Sun, C. Faustman, A. Senecal, A.L. Wilkinson and H. Furr, *Meat Sci.*, 2001, **57**, 55.
13. J. Sherma and B. Fried, *Practical Thin-Layer Chromatography: A Multidisciplinary Approach*, CRC Press, Boca Raton, 1996.
14. T. Hayashi, A. Terao, S. Ueda and M. Namiki, *Agric. Biol. Chem.*, 1985, **49**, 3139.
15. J.M. Ames, A. Apriyantono and A. Arnoldi, *Food Chem.*, 1993, **46**, 121.
16. J.M. Ames and A. Apriyantono, *Food Chem.*, 1994, **50**, 289.
17. H. Sakurai, H. Endo, T. Ito and I.K., *Bull. Coll. Agr. & Vet. Med.*, 1990, 88.
18. S.H. Slight, M.S. Feather and B.J. Ortwerth, *Biochim. Biophys. Acta*, 1990, **1038**, 367.
19. S.H. Slight, M. Prabhakaram, D.B. Shin, M.S. Feather and B.J. Ortwerth, *Biochim. Biophys. Acta*, 1992, **1117**, 199.
20. A.N. Wijewickreme, D.D. Kitts and T.D. Durance, *J. Agric. Food Chem.*, 1997, **45**, 4577.
21. L. Royle, R.G. Bailey and J.M. Ames, *Food Chem.*, 1998, **62**, 425.
22. T. Nakayama, F. Hayase and H. Kato, *Agric. Biol. Chem.*, 1980, **44**, 1201.
23. M. Wang, Y. Jin, J. Li and C.-T. Ho, *J. Agric. Food Chem.*, 1999, **47**, 48.
24. U.D. Neue, *HPLC Columns: Theory, Technology, and Practice*, John Wiley, New York, 1997.
25. S.J. Ge and T.C. Lee, *J. Agric. Food Chem.*, 1996, **44**, 1053.
26. J.S. Elmore, D.S. Mottram, M. Enser and J.D. Wood, *Meat Sci.*, 2000, **55**, 149.
27. M.O. Lederer and M. Baumann, *Bioorg. Med. Chem.*, 2000, **8**, 115.
28. M.O. Lederer, C.M. Dreisbusch and R.M. Bundschuh, *Carbohydr. Res.*, 1997, **301**, 111.
29. D.R. Sell, *Mech. Ageing Dev.*, 1997, **95**, 81.
30. G.C. Yen and P.P. Hsieh, *J. Sci. Food Agric.*, 1995, **67**, 415.
31. J.M. Ames, A. Arnoldi, L. Bates and M. Negroni, *J. Agric. Food Chem.*, 1997, **45**, 1256.
32. E.A. Pastorello and C. Trambaioli, *J. Chromatogr. B*, 2001, **756**, 71.
33. M.C. Garcia, M.L. Marina and M. Torre, *J. Chromatogr. A*, 2000, **880**, 169.
34. N. Innocente, C. Corradini, C. Blecker and M. Paquot, *Int. Dairy J.*, 1998, **8**, 981.
35. P. Caessens, H. Gruppen, C.J. Slangen, S. Visser and A.G.J. Voragen, *J. Agric. Food Chem.*, 1999, **47**, 1856.
36. P. Resmini and L. Pellegrino, *Cereal Chem.*, 1994, **71**, 254.
37. J.A. Gerrard, S.E. Fayle and K.H. Sutton, *J. Agric. Food Chem.*, 1999, **47**, 1183.
38. R. Oliver and B. King in *HPLC of Macromolecules – a Practical Approach*, ed. R. Oliver, Oxford University Press, Oxford, 1998, 1.
39. I. Nicoletti, E. Cogliandro, C. Corradini and D. Corradini, *J. Liq. Chromatogr. Relat. Technol.*, 1997, **20**, 719.
40. L. Pizzoferrato, P. Manzi, V. Vivanti, I. Nicoletti, C. Corradini and E. Cogliandro, *J. Food Prot.*, 1998, **61**, 235.
41. P.C. Sadek, *The HPLC Solvent Guide*, John Wiley & Sons Inc., New York, 1996.
42. D. Baunsgaard, L. Norgaard and H.A. Godshall, *J. Agric. Food Chem.*, 2001, **49**, 1687.
43. T. Knerr, H. Lerche, M. Pischetsrieder and T. Severin, *J. Agric. Food Chem.*, 2001, **49**, 1966.
44. E. Ferrer, A. Alegria, G. Courtois and R. Farre, *J. Chromatogr. A*, 2000, **881**, 599.
45. F.J. Morales and A. Arnoldi, *Food Chem.*, 1999, **67**, 185.

46. V. Fogliano, S.M. Monti, T. Musella, G. Randazzo and A. Ritieni, *Food Chem.*, 1999, **66**, 293.
47. Y. Al-Abed, D. Callaway, A. Kapurniotu, T. Holak, W. Voelter and R. Bucala, *Pol. J. Chem.*, 1999, **73**, 117.
48. J.P. Yuan and F. Chen, *Food Chem.*, 1999, **64**, 423.
49. T. Hofmann, *Z. Lebensm. Unters. Forsch. A*, 1998, **206**, 251.
50. M. Pischetsrieder, C. Schoetter and T. Severin, *J. Agric. Food Chem.*, 1998, **46**, 928.
51. M. Portero Otin, R. Pamplona, M.J. Bellmunt, M. Bergua, R.H. Nagaraj and J. Prat, *Life Sci.*, 1996, **60**, 279.
52. M. Pischetsrieder, *J. Agric. Food Chem.*, 1996, **44**, 2081.
53. V.M. Hill, D.A. Ledward and J.M. Ames, *J. Agric. Food Chem.*, 1996, **44**, 594.
54. A. Tirelli and L. Pellegrino, *Ital. J. Food Sci.*, 1995, **7**, 379.
55. M.A.E. Johansson, L. Fredholm, I. Bjerne and M. Jagerstad, *Food Chem. Toxicol.*, 1995, **33**, 993.
56. M. Zhu, D.C. Spink, B. Yan, S. Bank and A.P. DeCaprio, *Chem. Res. Toxicol.*, 1994, **7**, 551.
57. T. Yada, Y. Tabuchi, M. Fujii, T. Koh, Y. Tobimatsu, N. Hamasaki, Y. Sekiguchi, Y. Kato, M. Nakamura, M. Semma, M. Nishijima and Y. Ito, *Jpn. J. Toxicol. Environ. Health*, 1996, **42**, 417.
58. M. Franko, M. Sikovec, J. Kozar-Logar and D. Bicanic, *Anal. Sci.*, 2001, **17**, S515.
59. H.E. Indyk and E.L. Filonzi, *Aust. J. Dairy Technol.*, 2000, **55**, 99.
60. S. Vendrell-Pascuas, A.I. Castellote-Bargallo and M.C. Lopez-Sabater, *J. Chromatogr. A*, 2000, **881**, 591.
61. G.W. White, T. Katona and J.P. Zodda, *J. Pharm. Biomed. Anal.*, 1999, **20**, 905.
62. J.L. Casterline, C.J. Oles and Y.O. Ku, *J. AOAC Int.*, 1999, **82**, 759.
63. J. Onken and R.G. Berger, *Dtsch. Lebensm.-Rundsch.*, 1998, **94**, 287.
64. A. Brandenburg, *Sens. Actuator B-Chem.*, 1997, **39**, 266.
65. F.J. Morales and M. van Boekel, *Int. Dairy J.*, 1997, **7**, 675.
66. I.N. Shipanova, M.A. Glomb and R.H. Nagaraj, *Arch. Biochem. Biophys.*, 1997, **344**, 29.
67. M.E. Westwood and P.J. Thornalley, *J. Protein Chem.*, 1995, **14**, 359.
68. M.C. Mota, P. Carvalho, J.S. Ramalho, E. Cardoso, A.M. Gaspar and G. Abreu, *Int. Ophthalmol.*, 1994, **18**, 187.
69. R.H. Nagaraj and V.M. Monnier, *Biochim. Biophys. Acta-Protein Struct. Molec. Enzym.*, 1995, **1253**, 75.
70. L. Pellegrino, P. Resmini, I. DeNoni and F. Masotti, *J. Dairy Sci.*, 1996, **79**, 725.
71. F. Vinale, S.M. Monti, B. Panunzi and V. Fogliano, *J. Agric. Food Chem.*, 1999, **47**, 4700.
72. S.M. Monti, A. Ritieni, G. Graziani, G. Randazzo, L. Mannina, A.L. Segre and V. Fogliano, *J. Agric. Food Chem.*, 1999, **47**, 1506.
73. K. Gartenmann and S. Kochhar, *J. Agric. Food Chem.*, 1999, **47**, 5068.

CHAPTER 6

Mass Spectrometry

1 Introduction

Mass spectrometry (MS) has been adopted by many researchers attempting to unravel the mysteries of Maillard chemistry and has proven invaluable for the identification of a large number of the vast and varied products of the reaction. When a separation technique, such as HPLC or GC, is combined with MS, an extremely powerful analytical technique results. As these so-called hyphenated techniques have evolved, they have become some of the most widely used tools in our field, especially for the identification of the various aroma and flavour compounds present in foods.

2 How does MS Work?

Very simply, MS is a technique which involves the production of charged molecules, known as ions, which are then separated using electric and/or magnetic fields, according to their mass:charge (m/z) ratio (Figure 6.1).[1] Various methods

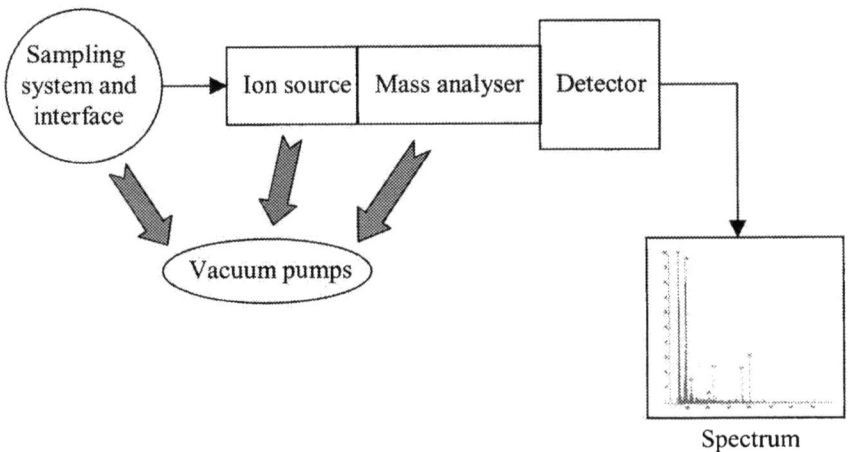

Figure 6.1 *Schematic of a MS instrument*

for ionising and separating sample ions are commercially available, many of which are discussed below. Once the sample has become ionised and its components separated, the ions pass into a detector. The MS process is performed under low pressure in order to minimise ion–molecule collisions prior to detection, since this may result in the loss of charge, preventing the target molecules from reaching the detector. When the ions strike the detector surface, they transfer their charge, generating a signal that is subsequently amplified, resulting in a current proportional to ion abundance. The relative abundance of each ion is then displayed as a plot of ion abundance against the m/z ratio and, if the ion is predominantly singly charged, the m/z ratio represents the mass of the ion.

3   Ionisation Techniques

Various methods are available for the production of ions that can then be accelerated through the analyser and into the detector. Those that have been used for the analysis of Maillard reaction systems include electron impact (EI) ionisation, chemical ionisation (CI), fast atom bombardment (FAB), electrospray ionisation (ESI) and matrix-assisted laser desorption ionisation (MALDI).

Electron Impact (EI)

EI is the most widely used ionisation technique in mass spectrometry. Ionisation occurs when a small amount of gaseous sample is bombarded by a beam of high-energy electrons. When the high-energy beam impacts upon a sample molecule, an electron is dislodged from the sample creating a singly charged radical cation, known as the molecular ion. The energy transferred from the electron to the sample is considerably greater than that required to form the radical cation. The remaining energy, therefore, causes the molecular ion to disintegrate, creating a number of smaller fragments – some of which retain a positive charge and some of which are neutral (Figure 6.2). Exactly how the molecule fragments, at any

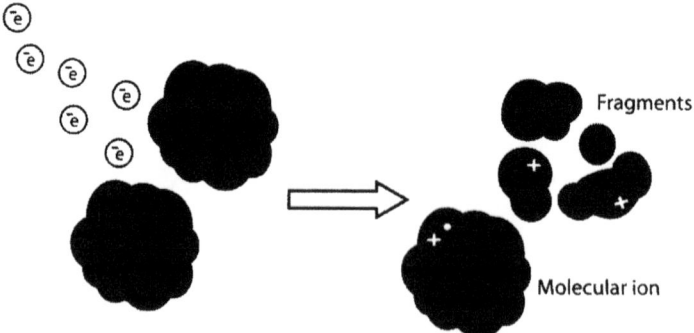

Figure 6.2   *Schematic representation of EI process involving the bombardment of sample molecules with a high-energy electron beam, ultimately producing a characteristic fragmentation pattern. Only positively charged fragments are detected*

given electron energy, is characteristic for that molecule. The resulting fragmentation pattern provides useful structural information that assists in the identification of the molecule.

Following fragmentation, only the positively charged species pass out of the ionisation chamber, through a series of electrically charged focusing lenses and into the mass analyser. Meanwhile, the remaining fragments are removed by vacuum. This form of ionisation has been used for the analysis of a number of the many small, volatile compounds produced by the Maillard reaction.[2-4]

Chemical Ionisation (CI)

One of the limitations of EI ionisation is that few, if any, of the molecular ions produced survive the fragmentation process. It is therefore difficult to determine the molecular weight of the analyte, since the molecular ion is absent from the EI spectrum. CI, on the other hand, is a 'softer' ionisation technique, occurring at much lower energy than EI, leading to little fragmentation (Figure 6.3). CI is generally suitable for the analysis of molecules with a molecular weight of up to 1000 Da.

This ionisation technique uses gases such as methane, ammonia or carbon dioxide to absorb the initial ionising electron. As these ionised gas molecules are very unstable, they will readily transfer their extra energy to a sample molecule if the opportunity arises. To increase the likelihood of this, the pressure within the ionisation chamber is higher than that used for EI, thus increasing the chances of the ionised gas molecules colliding with the sample molecules and transferring

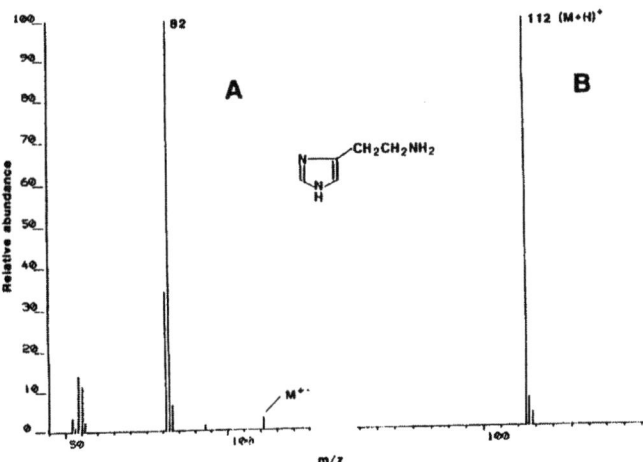

Figure 6.3 *The analysis of histamine (molecular weight 111 Da) using (A) EI and (B) CI. Note the abundance of the intact histamine molecule (M + H⁺) when CI is used*
(Reproduced with permission[1])

the excess energy of the extra proton. The ionised sample components then pass into the mass analyser for separation. Unlike EI, the molecular ion appears as a very strong peak in the mass spectrum and can be used to identify the molecular weight of the sample molecule. Few studies of the Maillard reaction have employed CI.

Fast Atom Bombardment (FAB)

CI and EI ionisation methods require the sample to be introduced into the ionisation chamber in a gaseous state. The FAB ionisation technique, however, provides a means of analysing non-volatile compounds since the sample molecules can be introduced in solution. This is done by grinding the sample with a viscous liquid, such as glycerol, and placing a drop of the emulsion onto a suitable probe. The probe is then inserted into the ionisation chamber where it is bombarded with heavy ions. As for EI, the FAB technique results in fragmentation and hence provides considerable structural information. The mass range for this ionisation technique varies with the type of mass analyser and detector used, but is generally up to 10 000 Da.

Electrospray Ionisation (ESI)

ESI is one of the most widely used, and fastest growing, ionisation techniques for the study of biomolecules. Large macromolecules, such as peptides and proteins, are typically inaccessible by the ionisation techniques mentioned above as, due to their size, these molecules are non-volatile and are thermally sensitive.

The electrospray process creates a fine aerosol of highly charged droplets, containing both solvent and analyte, in the presence of a strong electric field.[5] The loss of solvent results in a decrease in the size of the droplet, so that the charge eventually resides solely on the sample. The ions are then focused into an ion beam prior to entering the mass analyser.

The unique feature of ESI is its ability to produce ions which are multiply charged, the extent of which increases with molecular weight, bringing the resulting m/z ratio within the detectable limits of conventional mass spectrometers. For example, the addition of a single charge to a molecule, with a molecular weight of 30 000, results in a m/z ratio of 30 000, well outside the observable range of most quadrupole detectors. Using ESI, highly charged molecular ions can be formed, for example a 15+ ion, resulting in a m/z ratio of $30 000/15 = 2000$, which is within the range where mass spectrometers function well. The various charge state distributions can then be transformed to give a single peak corresponding to the molecular ion of the sample, as can be seen in Figure 6.4.

As fragmentation does not generally occur, electrospray ionisation is ideally suited to the analysis of protein or peptide mixtures, such as those produced by the Maillard reaction. Numerous studies of this type have appeared in the literature.[6-8]

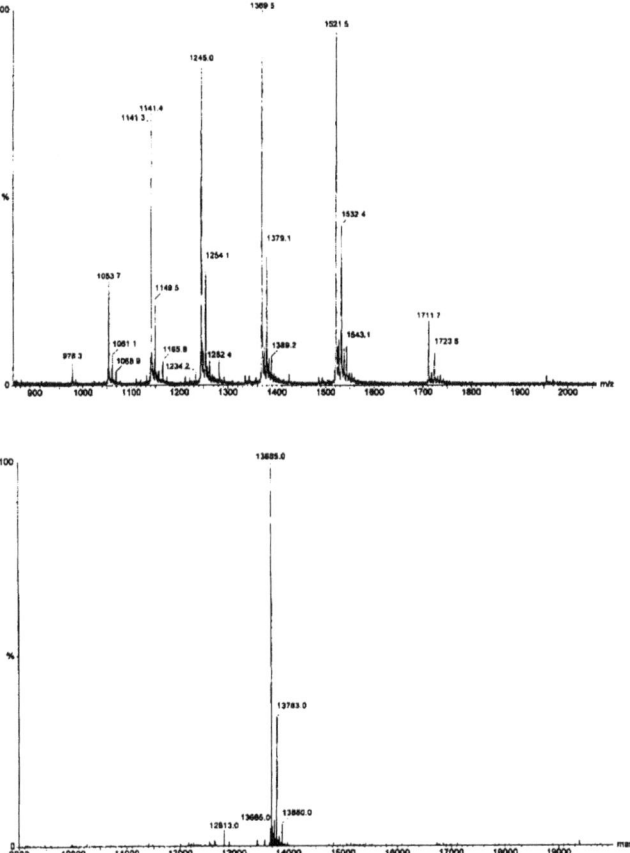

Figure 6.4 *Transformation of the mass spectrum of a multiply charged protein to a spectrum showing the molecular mass of the protein (molecular weight 13 685.0 Da)*

Matrix-assisted Laser Desorption Ionisation (MALDI)

As with ESI, MALDI produces intact molecular ions but, as these are primarily singly charged, they suffer from the low resolution achieved when working at high m/z. This form of ionisation involves mixing the sample molecules with a matrix, which acts as a type of intermediary. The matrix absorbs light from a pulsed laser beam, becomes ionised itself and then transfers its charge to the adjacent sample molecules, which can be up to 250 kDa in size. This form of ionisation has been one of the most commonly used in Maillard chemistry – specifically for the observation of peptide or protein-derived Maillard reaction products.[9-12]

4   Mass Analysers

Following ionisation, the ions pass into the mass analyser, where they are separated according to their m/z ratio. Various mass analysers are commercially available and are generally separated into two classes: magnetic analysers, such as the magnetic sector analyser, or non-magnetic analysers, such as the quadrupole, ion trap and time-of-flight mass analysers.

Magnetic Sector

In a magnetic sector analyser, ions are accelerated into the analyser where they are forced to follow a circular path by a magnetic field. For any given magnetic field strength, only ions with the appropriate m/z will follow the correct path to the detector. Other ions will adopt the wrong curvature and be deflected and lost, as shown in Figure 6.5. This may be used to monitor particular products of interest with specific m/z values, chosen to represent the target compound, and is known as selected-ion monitoring (SIM). SIM is useful if a particular compound is being monitored, such as some of the well-characterised marker compounds of the Maillard reaction. This, however, is at the expense of seeing the complete spectrum of reaction products. In this respect, it is similar to the immunological techniques described in Chapter 9. Alternatively, by altering the field strength, the entire m/z range can be scanned and a complete mass spectrum recorded.

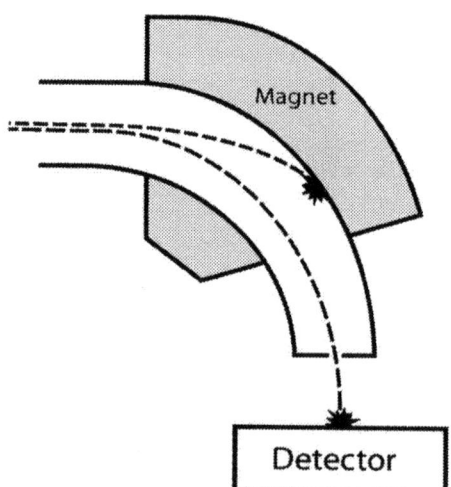

Figure 6.5   *Schematic representation of a magnetic sector analyser*

Quadrupole

One of the most commonly used analysers is the quadrupole mass analyser, since it is smaller, lighter and cheaper than most sector analysers. The pay-off for these attributes, however, is that the resolution is not as good as that achieved using a magnetic sector analyser.

The quadrupole analyser, shown in Figure 6.6, consists of four cylindrical quartz rods to which are applied both direct current and an oscillating radio frequency signal, with adjacent rods having opposite charge.[5] Only ions with a m/z ratio within a very narrow range, determined by the operator, can pass through the quadrupole and into the detector. Ions having a m/z outside this range travel along unstable trajectories, ultimately colliding with the quartz rods. These ions are not transmitted to the detector, hence the quadrupole analyser is also known as the quadrupole filter. Quadrupole mass analysers can also be operated in either SIM mode, or, by altering the voltage, for the scanning of a wide range of m/z values, known as SCAN mode.

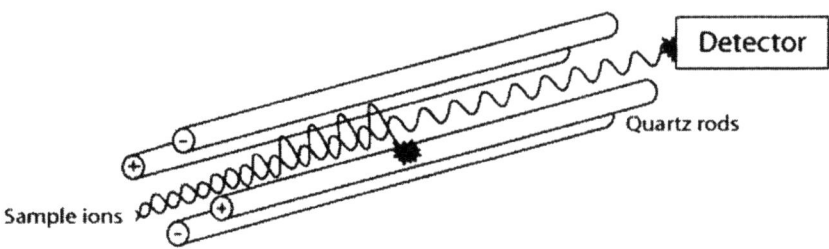

Figure 6.6 *Schematic representation of a quadrupole mass analyser. Adapted from Hofstadler et al.[5]*

Time-of-flight (TOF)

In the TOF analyser, all of the ions present in the sample are accelerated out of the ion source and are then left to drift down a long flight tube to the detector, with the lighter ions reaching the detector first and the heavier ions last. The m/z ratio can then be calculated from the measured flight time of the ions.

TOF analysers are predominantly used in association with MALDI, as they have a virtually unlimited m/z range and hence are able to resolve the large singly charged ions created by MALDI.

Ion Trap

The ion trap analyser involves either the injection of ions into the trap from an external ion source or the creation of ions within the trap by either CI or EI ionisation. The ions are then contained within the electrodes of the trap before being ejected in order of their m/z ratio (Figure 6.7). Surprisingly, the ion trap system has generally not been favoured by researchers in the Maillard field, despite increased sensitivity when compared with the quadrupole mass analyser.

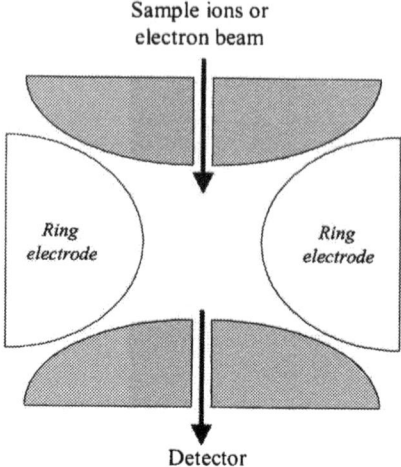

Sample ions or
electron beam

*Ring
electrode*

*Ring
electrode*

Detector

Figure 6.7   *Schematic representation of an ion trap mass analyser*

CASE STUDY 6.1 – The Analysis of Protein-derived Maillard Reaction Products

Traditional methods for the analysis of protein-derived Maillard products have generally involved the initial degradation of the protein backbone, by either acid or enzymatic hydrolysis, or by pyrolysis. These indirect methods of analysis increase the potential for introducing artefactual products that result from the degradation procedure. For this reason, the use of techniques that do not require sample degradation may prove to be of great importance. Examples of methods for the direct analysis of these products include electrophoresis (see Chapters 7 and 8), HPLC (see Chapter 5), MS and LC–MS.

Lapolla and co-workers have regularly employed MS in their studies of protein-derived Maillard reaction products.[13-15] For example, they demonstrated that MALDI-MS provides a useful means by which the extent of glycation can be measured. One such study involved the incubation of bovine serum albumin (BSA) with glucose.[16] As is shown in Figure 6.8, the observed mass was found to increase with time and was consistent with the consecutive addition of glucose molecules to the protein monomer. The maximum increase in molecular mass could be explained by the condensation of at least 51 glucose molecules. Presumably this involves the condensation of unreacted sugars with sugars that are already protein-bound, rather than all 51 sugars being bound to the protein directly. Similar results were obtained when this experiment was repeated with the protein ribonuclease A (RNAse A).[9]

Protein-derived Maillard reaction systems have also been studied using ESI.[8] Figure 6.9 (a) shows a single series of peaks with a mass of $13\,681 \pm 5$, consistent with the literature mass of the protein, which is 13 680 Da.[17] Following reaction with cyclotene, a volatile flavour compound commonly

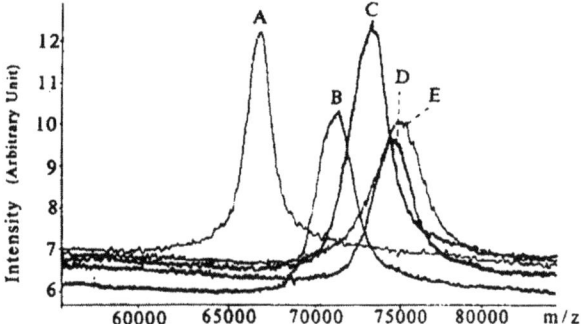

Figure 6.8 *MALDI spectra of BSA incubated with glucose, at 2 M concentration and 37 °C, recorded at different incubation times. (A) 0 days (molecular mass 66 429 Da), (B) 7 days (molecular mass 71 103 Da), (C) 14 days (molecular mass 73 099 Da), (D) 21 days (molecular mass 74 279 Da) and (E) 28 days (molecular mass 74 682 Da).*
(Reproduced with permission[14,16])

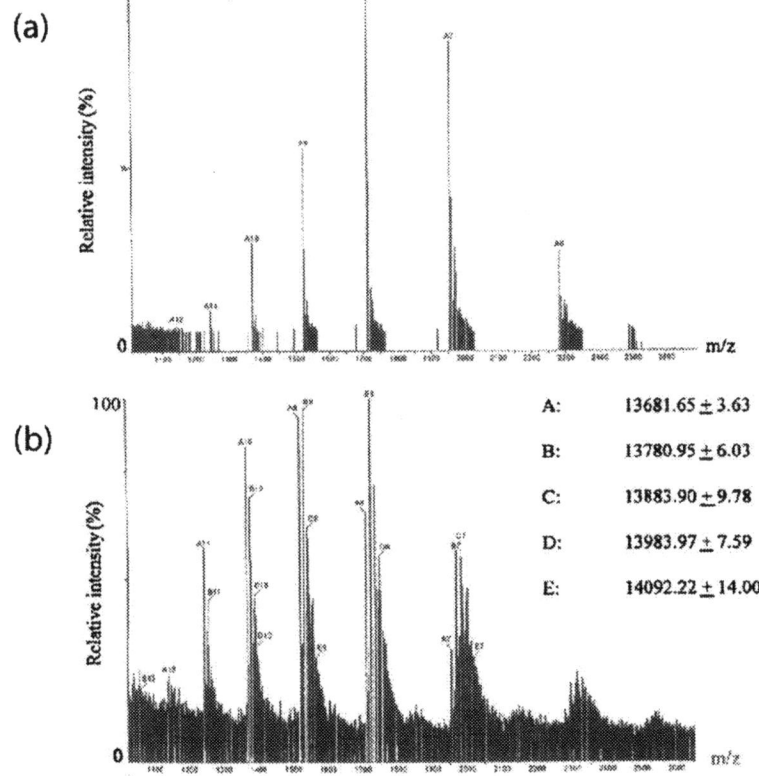

Figure 6.9 *ESI mass spectrum of (a) an incubated RNAse A control and (b) RNAse A incubated with cyclotene for 6 days at 37 °C[x]*

found in foods, five series of peaks can be seen (Figure 6.9 (b)). The first of these corresponds to unreacted protein. The remaining four are consistent with the successive addition of four molecules of cyclotene to the protein monomer. The addition of the first cyclotene molecule is believed to produce the Schiff base shown in Figure 6 '0. This is thought to be the intermediate in the formation of protein crosslinks *via* the Maillard reaction.

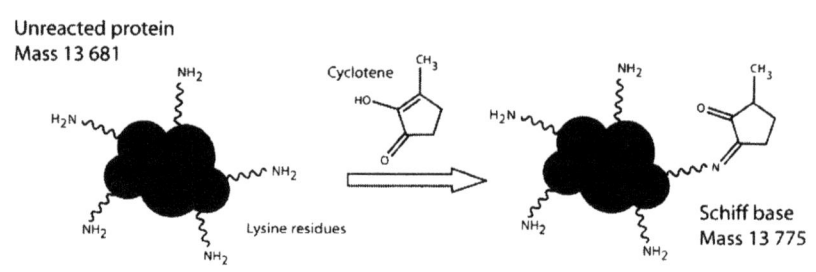

Figure 6.10   *Proposed formation of a Schiff base intermediate*

5   Hyphenated Techniques

Most investigations of the Maillard reaction involving MS have employed hyphenated techniques, where MS is used as a method of detection following separation by GC or LC. Tandem MS, in particular, has also demonstrated its worth for the study of Maillard chemistry. These techniques will be discussed below.

Gas Chromatography–Mass Spectrometry (GC–MS)

The combination of GC and MS in a single system results in a powerful technique – it is capable of separating a mixture into its individual components, identifying each component, as well as providing valuable structural information, all performed quickly and efficiently. In order to link a GC column to an MS instrument, the two instruments must be connected by an interface. The simplest interface may serve only to transfer the gaseous GC output directly into the ion source of the MS. Other interfaces, however, are capable of concentrating the GC output through the removal of much of the carrier gas, or can replace the carrier gas with one more suited to MS.[18]

A multitude of GC–MS studies have been published that explore the effects of the Maillard reaction, both in model systems and in actual foods. Recent examples of these include the analysis of polymeric products of the Maillard reaction,[10,19] as well as the identification of volatile products[20-23] of the reaction. (See Case Studies 3.1 and 5.1.)

Liquid Chromatography–Mass Spectrometry (LC–MS)

LC–MS is becoming increasingly popular with researchers in all fields of analysis, particularly with the advent of better interfaces that are capable of dealing with the large volumes of solvent produced during HPLC separation (Chapter 5).

LC–MS instruments have also become cheaper and smaller, making them more accessible to a greater number of laboratories.[18] This methodology has been employed by only a small number of researchers who study the Maillard reaction from a food perspective.[24-28] However, numerous researchers in the medical arena have adopted LC–MS technology in their studies of the Maillard reaction[29-32] and have developed methods which could easily be applied to the analysis of food systems.

CASE STUDY 6.2 – LC–MS Analysis of Lactuloselysine

As was discussed in Chapter 2, the Maillard reaction has been implicated in the loss of the nutritional quality of food, primarily due to the irreversible reaction of amino acids, particularly the amino acid lysine. In order to monitor the progression of the Maillard reaction during food processing, a number of products of the reaction have been identified as potentially useful marker compounds. The quantification of these compounds provides an indication of the nutritive damage that may have been caused by the Maillard reaction during food processing. In thermally-treated milk or milk products, the common markers include hydroxymethylfurfural (HMF), lactulose, *N*-ε-carboxymethyllysine (CML) and furosine.[33] A further compound which has proven to be of value in the analysis of milk products is lactuloselysine (*N*-ε-(1-deoxy-D-lactulosyl-1)-L-lysine), shown in Figure 6.11.

A variety of analytical methods have been developed for the analysis of lactuloselysine.[34-37] One method[28] utilises LC–MS for the quantification of protein-bound lactuloselysine.

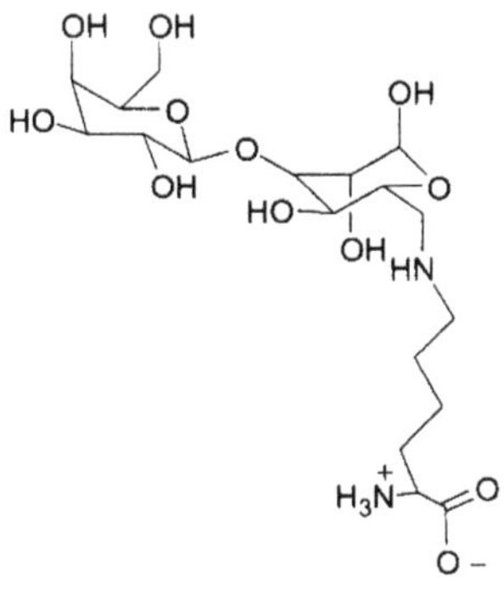

Figure 6.11 *Structure of lactuloselysine (N-ε-(1-deoxy-D-lactulosyl-1)-L-lysine)*

Having developed a method for the synthesis of lactuloselysine, Vinale *et al.*[28] confirmed the identity of the product by MALDI-TOF-MS, ESI-MS, UV and nuclear magnetic resonance (NMR) spectroscopy. LC–ESI-MS was used in SCAN mode to show the presence of lactuloselysine, indicating that the molecular ion had a mass of 471.0 Da, as shown in Figure 6.12. This mass was selected in SIM mode so as to increase the sensitivity of the analysis. These experiments were performed so that a calibration curve could be constructed, using different concentrations of lactuloselysine, and for the subsequent model studies.

The lactuloselysine contents of two model studies were measured. The first of these was a milk-like model system, comprising equimolar quantities of lysine monohydrochloride and lactose, which had been refluxed in distilled water. The second contained whey proteins, which had been isolated from commercially pasteurised skimmed milk, subsequently heated with lactose and enzymatically hydrolysed. They found that up to 50% of the lysine residues had been converted to lactuloselysine and that this value was in good agreement with previous work.[35]

The authors concluded that LC–MS was a powerful tool for the detection of Maillard reaction products, including those present in low concentrations.

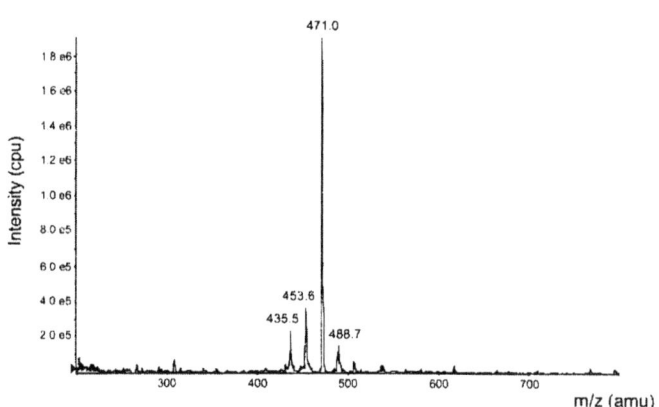

Figure 6.12   *ESI mass spectrum of lactuloselysine*
(Reproduced with permission[28])

Capillary Electrophoresis–Mass Spectrometry (CE–MS)

A further technique, which is rapidly gaining in popularity amongst researchers in the life sciences, is CE–MS.[38–41] After initial problems with the development of suitable interfaces, CE–MS is beginning to show utility for the analysis of complex mixtures and may prove to be of great value for the study of Maillard products in the future. CE methodology is described in Chapter 8.

Tandem Mass Spectrometry

A fourth hyphenated technique describes the combination of one MS mass analyser with another. This technique is often known as MS/MS, MS^2 or, if greater than two mass analysers are placed in sequence, MS^n. The first MS acts as a means of purifying the species of interest, while the second is used to analyse the purified ion and obtain structural information. Most MS/MS instruments consist of two mass analysers, such as quadrupole mass analysers, which are separated by a collision cell. There are a number of different modes of operation that can be utilised. For example, the first mass analyser can be set in the SIM mode, such that it transmits only ions of a specified m/z ratio. These ions then fragment in the collision cell and pass into the second mass analyser where the masses of the various fragments are scanned and recorded. Whereas traditional MS analysis would provide the fragment ions resulting from all of the compounds contained within a sample, this mode of operation provides information regarding the fragment ions that originate from a single molecular ion.

Fay *et al.*[42] demonstrated the benefits of tandem MS in their study of a pentose–glycine reaction system. The initial EI mass spectrum did not allow the unambiguous identification of the product. The second daughter spectrum, however, showing the fragment ions of the selected ion, definitively identified the product as 4-hydroxy-2,5-dimethyl-3(2*H*)-furanone.

Tandem MS techniques can also be used in conjunction with GC or HPLC. It is these techniques which have found the greatest use in the study of the Maillard reaction, especially in the analysis of food systems.[43-45]

6   Getting Started – in a Nutshell

The development of an MS method for the analysis of Maillard reaction systems is largely limited by the availability of instrumentation. Should you be lucky enough to have access to a wide range of instruments, then a number of questions need to be answered before a method can be developed. For instance, what do you know about the sample – how volatile is it? If it has low volatility, and must be heated, is it thermally stable? How pure is the sample? Are the target molecules small, or are they large peptides or proteins? What is the aim of the analysis – structural information, molecular mass or are both structural and mass information required? Are you looking for a specific compound, or do you require information on all of the products of the reaction?

Armed with the answers to these questions, the most appropriate ionisation technique and mass analyser can be chosen. With regard to a sample resulting from the Maillard reaction, it is possible that only the last of these questions can be answered. In this case, a trial and error approach may be necessary. For example, Ames *et al.*[46] found that an analysis of a coloured product of the Maillard reaction was unsuccessful when using EI, presumably due to the compound being insufficiently volatile. When FAB ionisation was utilised, the mass of the target compound was revealed, along with clues as to the structure of the compound.

In most Maillard research that employs mass spectrometry, the limiting factor proves to be separation of the many components of the system prior to injection into the mass spectrometer. As separation technologies improve, and interfacing devices come of age, mass spectrometry is set to become a major analytical tool for those investigating the Maillard reaction of food.

7 Further Reading

F.A. Mellon, R. Self and J.R. Startin, *Mass Spectrometry of Natural Substances in Food*, Royal Society of Chemistry, Cambridge, 2000.

M. McMaster and C. McMaster, *GC/MS – A Practical Users Guide*, Wiley-VCH, New York, 1998.

J.R. Chapman, *Mass Spectrometry of Proteins and Peptides*, Humana, New Jersey, 2000.

J.R. Chapman, *Practical Organic Mass Spectrometry: A Guide for Chemical and Biochemical Analysis*, Wiley, New York, 1995.

R.A.W. Johnstone and M.E. Rose, *Mass Spectrometry for Chemists and Biochemists*, 2nd ed., Cambridge University Press, Cambridge, 1996.

8 References

1. J.R. Chapman, *Mass Spectrometry of Proteins and Peptides*, Humana, New Jersey, 2000.
2. I. Cutzach, P. Chatonnet, R. Henry and D. Dubourdieu, *J. Agric. Food Chem.*, 1997, **45**, 2217.
3. V.M. Hill, N.S. Isaacs, D.A. Ledward and J.M. Ames, *J. Agric. Food Chem.*, 1999, **47**, 3675.
4. W.L.P. Bredie, D.S. Mottram and R.C.E. Guy, *J. Agric. Food Chem.*, 1998, **46**, 1479.
5. S.A. Hofstadler, R. Bakhtiar and R.D. Smith, *J. Chem. Educ.*, 1996, **73**, A82.
6. F.K. Yeboah, I. Alli, V.A. Yaylayan, Y. Konishi and P. Stefanowicz, *J. Agric. Food Chem.*, 2000, **48**, 2766.
7. D. Favratto, R. Seraglia, P. Traldi, O. Curcuruto and M. Hamdan, *Org. Mass Spectrom.*, 1994, **29**, 526.
8. J.A. Gerrard, S.E. Fayle and K.H. Sutton, *J. Agric. Food Chem.*, 1999, **47**, 1183.
9. A. Lapolla, L. Baldo, R. Aronica, C. Gerhardinger, D. Fedele, G. Elli, R. Seraglia, S. Catinella and P. Traldi, *Biol. Mass Spectrom.*, 1994, **23**, 241.
10. R. Tressl, G.T. Wondrak, L.-A. Garbe, R.-P. Krueger and D. Rewicki, *J. Agric. Food Chem.*, 1998, **46**, 1765.
11. U. Tagami, S. Akashi, T. Mizukoshi, E. Suzuki and K. Hirayama, *J. Mass Spectrom.*, 2000, **35**, 131.
12. H.J. Kim, J. Leszyk and I.A. Taub, *J. Agric. Food Chem.*, 1997, **45**, 2158.
13. A. Lapolla, D. Fedele, M. Plebani, M. Garbeglio, R. Seraglia, D. D'Alpaos, C.N. Aric and P. Traldi, *Rapid Commun. Mass Spectrom.*, 1999, **13**, 8.
14. P. Traldi, A. Lapolla, R. Seraglia, S. Catinella, M. Dalpaos, R. Aronica and D. Fedele, *Microchem. J.*, 1996, **54**, 218.
15. A. Lapolla, D. Fedele, R. Aronica, L. Baldo, M. D'Alpaos, R. Seraglia and P. Traldi, *Rapid Commun. Mass Spectrom.*, 1996, **10**, 1512.
16. A. Lapolla, C. Gerhardinger, L. Baldo, D. Fedele, A. Keane, R. Seraglia, S. Catinella and P. Traldi, *Biochim. Biophys. Acta*, 1993, **1225**, 33.
17. M. Jullien, M.-P. Crosio and S. Baudet-Nessler, *J. Mol. Biol.*, 1992, **228**, 243.

18. M. McMaster and C. McMaster, *GC/MS – A Practical Users Guide*, Wiley-VCH, New York, 1998.
19. S.M. Rogacheva, M.J. Kuntcheva, T.D. Obretenov and G. Vernin in *The Maillard Reaction in Foods and Medicine*, eds. J. O'Brien, H.E. Nursten, M.J.C. Crabbe and J. M. Ames, Royal Society of Chemistry, Cambridge, 1998, 89.
20. C.M. Wu, Z.Y. Wang and Q.H. Wu, *J. Agric. Food Chem.*, 2000, **48**, 2438.
21. F.B. Whitfield and D.S. Mottram, *J. Agric. Food Chem.*, 1999, **47**, 1626.
22. S. Muresan, M. Eillebrecht, T.C. de Rijk, H.G. de Jonge, T. Leguijt and H.H. Nijhuis, *Food Chem.*, 2000, **68**, 167.
23. J.K. Parker, G.M.E. Hassell, D.S. Mottram and R.C.E. Guy, *J. Agric. Food Chem.*, 2000, **48**, 3497.
24. S.M. Monti, R.C. Borrelli, A. Ritieni and V. Fogliano, *J. Agric. Food Chem.*, 2000, **48**, 1041.
25. S.M. Monti, A. Ritieni, G. Graziani, G. Randazzo, L. Mannina, A.L. Segre and V. Fogliano, *J. Agric. Food Chem.*, 1999, **47**, 1506.
26. T. Hofmann, *J. Agric. Food Chem.*, 1998, **46**, 3902.
27. T. Hofmann, *Carbohydr. Res.*, 1998, **313**, 203.
28. F. Vinale, S.M. Monti, B. Panunzi and V. Fogliano, *J. Agric. Food Chem.*, 1999, **47**, 4700.
29. P.A. Harmon, W. Yin, W.E. Bowen, R.J. Tyrrell and R.A. Reed, *J. Pharm. Sci.*, 2000, **89**, 920.
30. M.O. Lederer and H.P. Buhler, *Bioorg. Med. Chem.*, 1999, **7**, 1081.
31. H. Odani, Y. Matsumoto, T. Shinzato, J. Usami and K. Maeda, *J. Chromatogr., B*, 1999, **731**, 131.
32. S. Li, T.W. Patapoff, D. Overcashier, C. Hsu, T.H. Nguyen and R.T. Borchardt, *J. Pharm. Sci.*, 1996, **85**, 873.
33. H.F. Erbersdobler, J. Hartkopf, H. Kayser and A. Ruttkat, *ACS Symp. Ser.*, 1996, **631**, 45.
34. L. Mao, Z. Chen, W. Li and L. Jin, *Zhejiang Nongye Daxue Xuebao*, 1997, **23**, 437.
35. T. Henle, H. Walter and H. Klostermeyer, *Z. Lebensm.-Unters. Forsch.*, 1991, **193**, 119.
36. V. Fogliano, S.M. Monti, A. Ritieni, C. Marchisano, G. Peluso and G. Randazzo, *Food Chem.*, 1997, **58**, 53.
37. R. Pizzano, M.A. Nicolai, R. Siciliano and F. Addeo, *J. Agric. Food Chem.*, 1998, **46**, 5373.
38. K.B. Presto Elgstoen, J.Y. Zhao, J.F. Anacleto and E. Jellum, *J. Chromatogr., A*, 2001, **914**, 265.
39. A.B. Wey, J. Caslavska and W. Thormann, *J. Chromatogr., A*, 2000, **895**, 133.
40. M. Serwe and G.A. Ross, *LC-GC*, 2000, **18**, 46.
41. T. Soga and D.N. Heiger, *Anal. Chem.*, 2000, **72**, 1236.
42. L.B. Fay, I. Blank and C. Cerny in *Flavour Science: Recent Developments*, eds. A.J. Taylor and D.S. Mottram, Royal Society of Chemistry, Cambridge, 1996, 271.
43. L.B. Fay, *Analysis*, 1998, **26**, M28.
44. D. Molle, F. Morgan, S. Bouhallab and J. Leonil, *Anal. Biochem.*, 1998, **259**, 152.
45. M. Amranihemaimi, C. Cerny and L.B. Fay, *J. Agric. Food Chem.*, 1995, **43**, 2818.
46. J.M. Ames, A. Apriyantono and A. Arnoldi, *Food Chem.*, 1993, **46**, 121.

CHAPTER 7

Electrophoresis

1 Introduction

Despite prolific use of electrophoresis for the analysis of proteins throughout biochemistry, electrophoretic techniques have been strikingly under-used in the field of food science. Areas where electrophoresis has proven to be of particular value in the food arena include wheat research and the meat and dairy industries.[1-5] In each of these areas, electrophoresis is routinely used for evaluating protein composition. This information is vital for the determination of the end-use of the raw materials, as the protein content and specific composition can have a profound effect on the quality of the final product. For instance, wheat flours with a high protein content, particularly those containing specific combinations of proteins, are ideally suited to the production of high quality breads, whereas flours with low protein content are better suited to the production of high quality biscuits.[6] Electrophoresis provides a useful means for the rapid determination of protein composition and, therefore, an indication as to the best use for the flour, without the need for test-baking.

Although electrophoresis has found use in the study of the Maillard reaction from a medical perspective,[7-10] it has yet to be used routinely by those employed in the food arena. Perhaps due to the historical focus on small molecules, the few studies of the Maillard reaction of food proteins have generally involved the analysis of the constituent amino acids, following an acid digest.[11] As a number of the products of the Maillard reaction are thought to be acid-labile, the use of an acid digest increases the ease of analysis, since any acid-labile products will no longer be present. However, the potential risk of creating artefactual results at low pH is also increased. Alternatively, the direct analysis of the intact protein, for which electrophoresis is ideally suited, avoids this problem. Moreover, the results give a better representation of actual Maillard products as the possibility of further reaction during analysis has been minimised.

Electrophoresis is a relatively simple, yet highly sensitive, separation technique that is invaluable when exploring the physicochemical properties of proteins. Electrophoretic separation exploits the fact that macromolecules, such as proteins, bear an intrinsic net charge. When an electric field is applied to a protein mixture, the anions migrate towards the anode and the cations toward the cathode, at a rate determined by their mass:charge ratio (see Figure 7.1).

A variety of different support materials, such as paper, agarose, silica and starch, can be used for electrophoretic separation. The most popular gel support for the separation of proteins is polyacrylamide, since it is chemically inert and is readily formed by the polymerisation of acrylamide with the crosslinker N,N'-methylenebisacrylamide, commonly known as bisacrylamide, as shown in Figure 7.2. The versatility of polyacrylamide electrophoresis arises largely from the ability to control the porosity of the gel, which can be altered over a wide range, enabling the separation of molecules of vastly different sizes. This is done simply by varying the ratio of acrylamide to bisacrylamide. In this manner, polypeptides as small as 1 kDa can be separated from larger proteins, which can be as large as 300 kDa.

The polyacrylamide gel can be manipulated so that it serves either as a molecular sieve, for separations based on size, or simply as a structural support, for separation by charge. For separation by size, molecules which are small, compared with the pores in the gel, are able to travel through the gel with ease, whereas larger molecules squeeze through the pores of the gel with more difficulty, thereby reducing the speed of travel through the gel. The largest molecules are essentially immobile as they are unable to enter the gel matrix and these

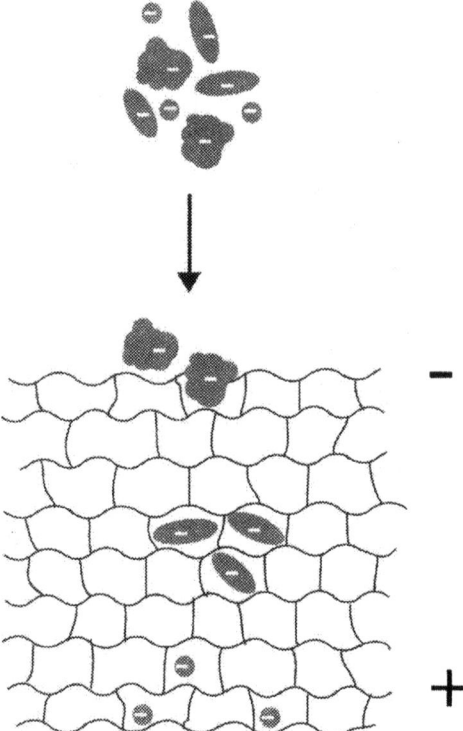

Figure 7.1 *Schematic of separation of proteins by size within a polyacrylamide gel matrix*

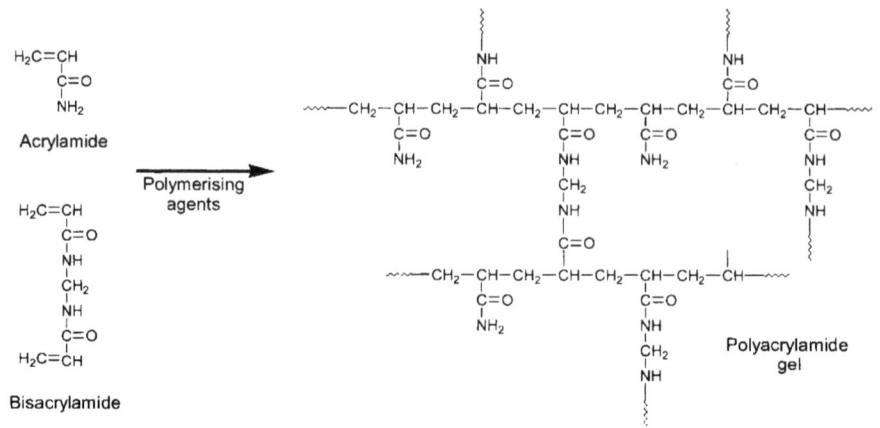

Figure 7.2 *The formation of a polyacrylamide gel matrix from acrylamide monomer and bisacrylamide*

molecules can often be found sitting at the upper edge of the gel, almost exactly where they were initially loaded. When the basis of separation is due to variations in charge between proteins, a pore size is chosen which will not inhibit even the largest of proteins. In this instance the mobility of the proteins is determined solely by their intrinsic charge.

Protein mixtures are analysed by loading onto the top of the polyacrylamide gel, which is typically in the form of a vertical slab, and electrophoresed by applying a current across the electrodes. The separated proteins can then be visualised by staining them with a dye, as discussed later in this chapter.

The term electrophoresis includes a vast, and continually growing, range of specialised methodologies and protocols, many of which are beyond the scope of this book. This chapter will instead serve purely as an introduction to the potential of electrophoresis, particularly discussing techniques that have already found favour with researchers working in the field of Maillard chemistry. Numerous informative texts are available which provide more in-depth discussions of both the fundamental theory and practical aspects of electrophoresis. Some suggested texts are listed at the end of this chapter.

2   Separation by Size

Evidence suggests that the Maillard reaction can lead to the formation of protein aggregates.[12,13] These crosslinked proteins may not only result from the direct reaction of proteins with sugars, but also through reaction with any of the low molecular weight compounds produced by the Maillard reaction.

The formation of protein crosslinks *via* the Maillard reaction is important in food systems, where the crosslinking can affect the structure and, hence, the function of the protein. For example, food proteins crosslinked by the Maillard reaction have been found to demonstrate improved gelling properties, with breakstrengths much higher than those of monomeric protein.[14] (See Case Study

7.2.) Other functional properties that may be affected include texture, solubility, emulsification, viscosity and enzyme activity. Crosslinking can also result in decreased nutritional quality of food products through the destruction of essential amino acids, particularly the amino acid lysine, and a decrease in digestibility of aggregated protein.[15]

Since the Maillard reaction is very complex, and the chemistry of macromolecules is intrinsically difficult to study, little is known about the molecular details associated with protein crosslinking and aggregation. More research is required so that the fundamental chemistry of the crosslinking mechanism can be unveiled. It may then be possible to manipulate the reaction for positive effect, by the deliberate introduction or removal of covalent crosslinks. A thorough understanding of the mechanisms involved in the formation of these crosslinks may, therefore, provide a tool for the improvement and maintenance of food quality. Electrophoresis provides a simple and convenient method by which to study such processes.

Intermolecular crosslinking of proteins is readily detected since this process results in multimeric proteins, with a consequent change in mass, as depicted in Figure 7.3. This change in mass can be observed after an electrophoretic protocol in which separation is based on differences in mass. The electrophoretic technique most appropriate for the study of protein crosslinking as a result of the Maillard reaction is sodium dodecyl sulfate–polyacrylamide gel electrophoresis (SDS–PAGE).

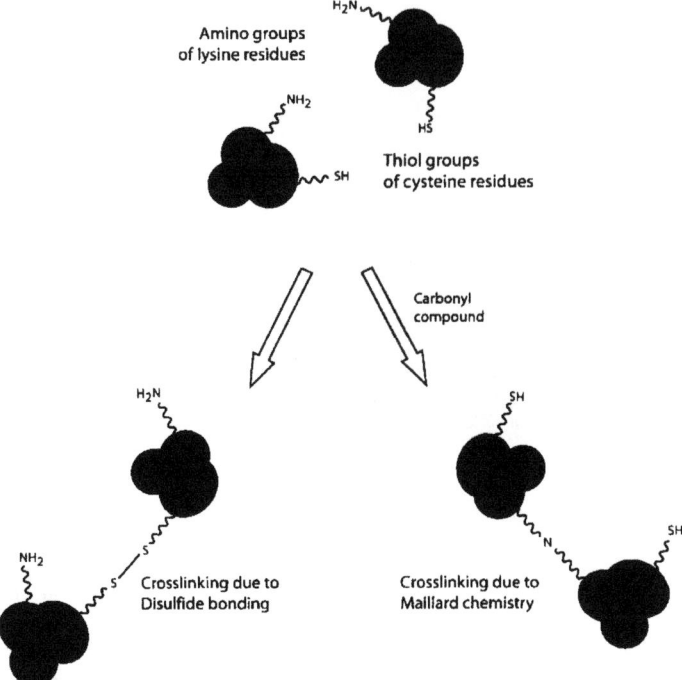

Figure 7.3 *Intermolecular crosslinking of proteins resulting from either disulfide bonding or Maillard chemistry*

SDS–PAGE is one of the most commonly used systems for the routine analysis of proteins. SDS acts as a denaturing agent, which increases the solubility of the sample and confers a uniform negative charge density upon the resulting protein–detergent complexes. As a result, the proteins migrate through the gel matrix at a rate determined by their molecular size rather than by the charge conferred by their amino acid composition, since this is masked by the overall negative charge of the SDS. Electrophoresis in a reductive environment, that is, in the presence of a reducing agent such as mercaptoethanol or dithiothreitol, ensures that all disulfide bonds have been broken. Changes in mass detected under these conditions must therefore be attributed to alternative crosslinking methods, such as the Maillard reaction.

SDS–PAGE can provide information regarding the size of the protein, an estimation of its purity, a comparison of the polypeptide composition of samples before and after reaction with a reducing sugar, and protein quantitation. An example of the power of this methodology is given in Case Study 7.1.

CASE STUDY 7.1 – The Analysis of the Maillard Reaction of Food Proteins by SDS–PAGE

SDS–PAGE has often been used for the analysis of Maillard reaction products in the medical arena[16-19] but has not been widely adopted by those studying

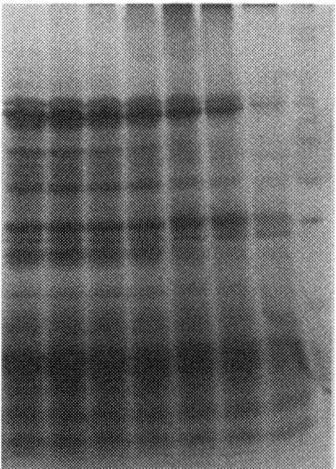

←— Aggregates

1 2 3 4 5 6 7 8

Figure 7.4   *SDS–PAGE gel showing the reaction of the albumin and globulin fraction of wheat proteins with glyceraldehyde in the presence of the reducing agent mercaptoethanol. Lanes (from left to right): 1, fresh protein only control; 2–7, albumin and globulin protein fractions incubated with glyceraldehyde for 2, 4, 8, 12, 24 and 48 hours respectively; 8, markers. Protein aggregation, the result of Maillard crosslinking, is clearly visible in lanes 2- 7*
(Reproduced with permission[22])

the reaction from a food perspective.[20,21] A recent study has shown that this methodology is readily applicable to proteinaceous Maillard products resulting from the reaction of wheat proteins with carbonyl compounds.[22]

Figure 7.4 depicts an SDS–PAGE gel in which proteins were extracted from dough and incubated with glyceraldehyde prior to analysis. The addition of glyceraldehyde resulted in the formation of crosslinked protein, which increased with time. This study demonstrates that gel electrophoresis provides a simple, yet useful, means of assessing this type of reaction.

3 Separation by Charge

Whereas SDS–PAGE separates proteins with respect to their size, other electrophoretic techniques, such as isoelectric focusing, separate proteins according to their overall charge. Hence, if the gel matrix contains large pores, which allow all of the proteins to move freely throughout the gel, only modifications in the net charge of the protein will alter its mobility in the electrophoretic gel. In the case of native gel electrophoresis, both the size and shape of the protein may contribute to differences in mobility in addition to differences in charge.

Isoelectric Focusing

An example of a method where separation is based on the charge of the protein is isoelectric focusing (IEF). IEF uses a pH gradient to separate proteins according to their isoelectric point (pI), which is the pH at which the net charge of the protein is zero. At this pH, its electrophoretic mobility will be zero (Figure 7.5). Therefore, a mixture of proteins, each with a different pI, will travel through the gel until each protein reaches a position in the gel where its pI equals the pH. The proteins are focused at a particular point along the pH gradient.[23]

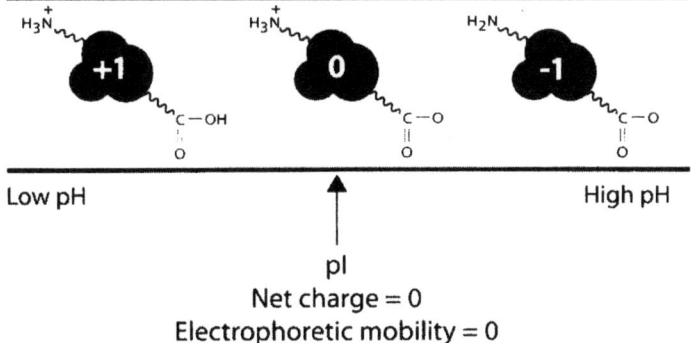

Figure 7.5 *Schematic showing the pI of a protein changing with pH. When at its pI, the net charge of the protein is zero and its electrophoretic mobility will also be zero*

The pH gradient used to separate protein mixtures can be created by the addition of amphoteric buffers, capable of acting both as an acid or a base, known as ampholytes, which have closely spaced pIs across a given pH range. When an electric field is applied, these compounds also migrate until their pI equals the pH, forming a pH gradient where the most acidic of the ampholytes is closest to the anode and the most basic is closest to the cathode.[23]

An alternative to carrier ampholytes, for the generation of a pH gradient, is the use of immobilised pH gradients (IPG). The pH gradient described above was achieved with the use of amphoteric compounds that were able to migrate under an electric field. IPGs, on the other hand, are created by adding a range of non-amphoteric compounds to the gel solution. These compounds are generally weak acids and bases, which have a double bond at one end, similar to that of acrylamide, shown in Figure 7.6. They can therefore be polymerised within the polyacrylamide gel and are fixed in place. IPGs have become the method of choice for many researchers due to their ease of use, excellent resolution and reproducibility.[23]

IEF can readily resolve proteins which differ by one net charge, providing a powerful means of separating complicated mixtures such as those created by the Maillard reaction. IEF methodology has rarely been utilised by researchers in this field.

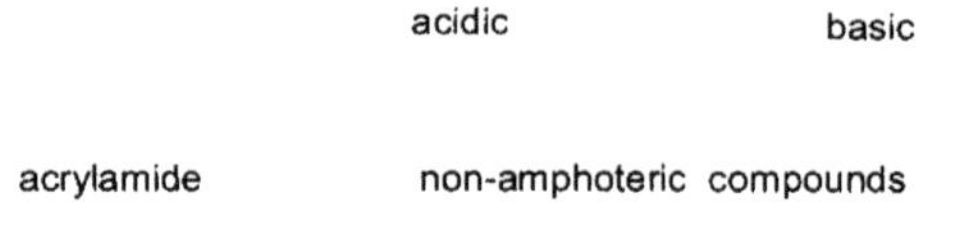

Figure 7.6   *Comparison of the structure of acrylamide with typical non-amphoteric compounds, used to create IPGs*

Native Gel Electrophoresis

Native gel electrophoresis separates proteins by a combination of their size and charge. Proteins, which are of approximately the same size, and hence can not be separated by SDS–PAGE, may prove easier to resolve using this technique. Native gel electrophoresis has also not attracted the attention of those studying Maillard chemistry. A useful description of this technique, including references for specific protocols, can be found in the opening chapter of Hames.[24]

CASE STUDY 7.2 – Maillard Chemistry: A Useful Tool for Food Processors?

In recent years, the Maillard reaction has attracted increasing attention as a tool for improving the functional properties of proteins.[20,25-27] Two properties which are known to benefit from Maillard chemistry are gelation[14] and emulsification.[28]

Gels are developed in order to create a food product with a particular consistency or texture, or to provide physical stability such that the product can withstand handling and slicing. Several globular proteins are known to form gels irreversibly as a result of the Maillard reaction, for example, egg white, soya isolate and bovine serum albumin.[29] The gelation process is believed to involve the unfolding of the protein followed by the formation of covalent crosslinks, salt bridges and/or hydrophobic interactions.[30] It is thought that the Maillard reaction improves the gelling properties of the protein by the formation of non-disulfide covalent crosslinks, which are readily observable by SDS–PAGE.

Handa and Kuroda[26] have recently employed electrophoretic techniques in their study of the application of Maillard chemistry to the improvement of the functional properties of sugar-preserved dried egg white. Commercially separated egg white was spray dried with 0.4% (w/w) citric acid prior to reaction with 150 mg mL^{-1} glucose for up to twelve days, at a temperature of 55 °C. Changes in charge and size were monitored by native PAGE and SDS–PAGE respectively.

The native PAGE gel (Figure 7.7) shows the migration of the ovalbumin and ovotransferrin bands towards the anode, indicating an increase in negative

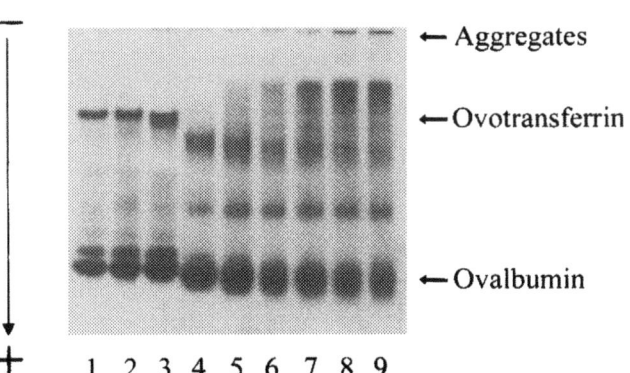

← Aggregates

←Ovotransferrin

←Ovalbumin

1 2 3 4 5 6 7 8 9

Figure 7.7 *Native PAGE gel showing the reaction of dried egg white with glucose. Lanes (from left to right): 1, non-heated desugared dried egg white; 2, dried egg white heated for 12 days; 3, non-heated dried egg white; 4–9, dried egg white heated with glucose for 2, 4, 6, 8, 10 and 12 days respectively*
(Reproduced with permission[26])

charge with increasing reaction time. This is consistent with the attachment of a glucose molecule to the amino group of a lysine residue that has also been described by other researchers (see Case Study 6.1).[22,31] The concomitant formation of protein aggregates was observed by SDS–PAGE, as shown in Figure 7.8. Gels were run under both reducing and non-reducing conditions. When the reducing agent mercaptoethanol was present, the amount of aggregate formation decreased only slightly, indicating that not all of the cross-

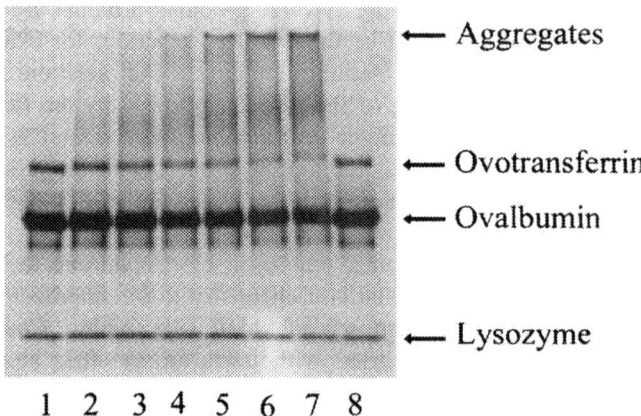

Figure 7.8 *SDS–PAGE gel showing the reaction of dried egg white with glucose in the presence of the reducing agent mercaptoethanol. Lanes (from left to right): 1, non-heated desugared dried egg white; 2–7, dried egg white heated with glucose for 2, 4, 6, 8, 10 and 12 days respectively; 8, non-heated dried egg white*
(Reproduced with permission[26])

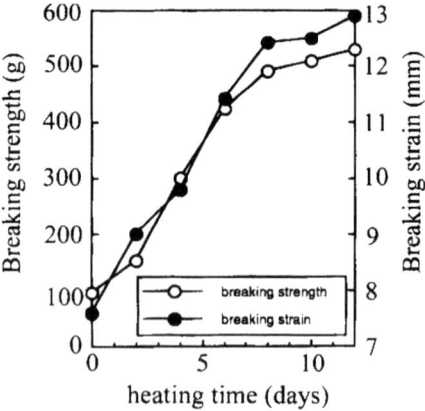

Figure 7.9 *Changes in breaking strength and breaking strain of heat-induced gels from dried egg white heated with glucose*
(Reproduced with permission[26])

linking was due to disulfide bonding. Non-disulfide protein crosslinks, such as those produced by the Maillard reaction, were also being formed. No crosslinked aggregates were observed in the samples that did not contain sugar.

Samples of dried egg white that had been incubated with glucose for up to twelve days were used to prepare gels. As can be seen in Figure 7.9, the break strengths of the gels were found to increase with increasing time of incubation. This suggests that the increase in the gelling properties of the glucose–dried egg white reaction mixture may be attributed to an increase in protein crosslinks as a result of the Maillard reaction.

4 Separation by Size and Charge – a Second Dimension

The co-migration of Maillard reaction products is a potential problem when analysis is performed by one-dimensional electrophoresis (1-DE). This may not only mask the true complexity of the sample, but also mask the apparent amount, or changes in the amount, of components present in any band. For this reason two-dimensional electrophoresis (2-DE) methods, commonly used by researchers in the proteomics field, may prove to be of particular importance.

2-DE methods, which combine IEF with SDS–PAGE, provide a powerful means of producing a gel that contains very highly resolved proteins. The sample is initially separated by IEF using a gel strip. The strip is then inserted into the top of an SDS–PAGE gel and electrophoresed in the second dimension,

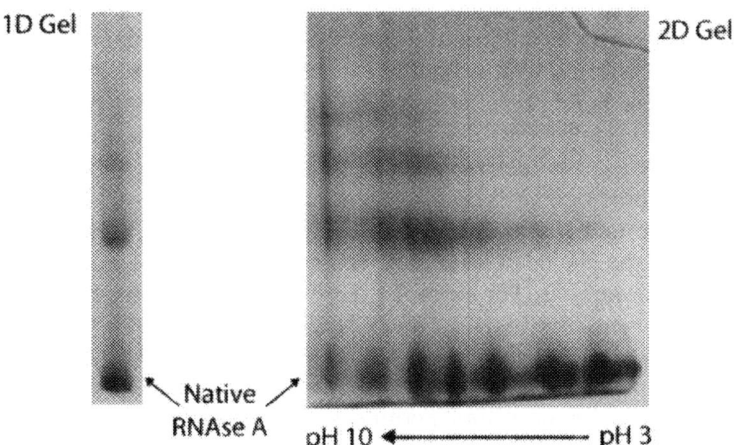

Figure 7.10 *1-DE and 2-DE gels (run under reducing conditions) showing a sample containing the protein RNAse A and the carbonyl compound cyclotene, heated at 37 °C for 32 days*[22]

perpendicular to the first. This combination allows the separation by pI in the horizontal direction followed by separation by size in the vertical direction.[23]

In the proteomics field, the resulting 2-DE gels can contain as many as 1000 well-resolved protein spots; the technique thus seems highly appropriate for the analysis of the complex mixtures of Maillard reaction products resulting from the reaction of sugars with proteins. The analysis of a Maillard reaction mixture by 2-DE, using a pH gradient of 3–10, is shown in Figure 7.10.[22] Comparison of the 2-DE gel with the 1-DE SDS–PAGE gel of the same sample, clearly demonstrates the complexity of Maillard reaction samples which is not obvious when analysis is performed by the 1-DE technique alone. 2-DE techniques may therefore prove to be immensely powerful in our search for a greater understanding of the Maillard reaction in food systems, particularly as methods for removing resolved proteins from gels, for subsequent identification and analysis (see Section 6), become more sophisticated.

5   Visualisation of Proteins

Having separated a protein mixture by either 1-DE or 2-DE, the resolved proteins can be visualised using any of a wide variety of techniques.[32] There is a vast number of stains available; the most commonly used are the silver staining techniques[33] and staining with the organic dyes Coomassie brilliant blue R250 and G250.[34]

Protein stains have varying sensitivity limits. Silver stain is typically the most sensitive stain available. The detection limit of silver staining can be improved further by using it in combination with any of the other stains, *e.g.* Coomassie brilliant blue R250. The stains with lower sensitivity are generally much easier and quicker to use in comparison with silver staining, although SYPRO ruby, a ready-to-use staining solution, provides comparable sensitivity.[23]

Fluorescent staining techniques are also available.[35] These generally involve the labelling of the proteins with a fluorescent marker, such as SYPRO orange and SYPRO red, either before or after an electrophoretic separation. Autoradiography is used for the visualisation of radio-labelled proteins. However, if autoradiography is to be used, it is preferable not to stain the gel; instead, the gel should be soaked in an autoradiography reagent.[36]

If proteins are to be eluted from the gel (see Section 6) for direct chemical characterisation, the choice of stain must be made with care, because not all stains allow subsequent analysis. For example, the Coomassie brilliant blue G250 stain can not be used on proteins for which N-terminal sequencing data are to be acquired. If N-sequencing is required, Coomassie brilliant blue R250 should be used to stain the gel. For subsequent mass spectrometric detection (described in Chapter 6), SYPRO ruby or, if time permits, silver nitrate are the stains of choice.

Having stained the gel, the relative quantities of the resolved protein components can be assessed. Although a rough estimate can be obtained by visually inspecting the gel, this can also be achieved by imaging the polyacrylamide gel and calculating the relative amounts of each component present. However, as individual proteins can respond differently to the same stain due to differences in the amino acid composition or the presence of non-protein groups, such as carbohydrate – of particular concern to Maillard researchers – the resulting data should be considered a guide to the relative quantities rather than an absolute data set.[37] Quantification is most commonly achieved through the use of a densitometer that calculates the protein content by measuring the amount of light transmitted through the gel.

6 Beyond Electrophoresis – Further Analysis

Having separated a protein mixture by 2-DE, the resolved proteins can be transferred to a solid support. This procedure is known as electroblotting or Western blotting. Proteins can then be analysed further in order to identify and characterise the individual proteins.[16] The transfer of separated proteins from an SDS–PAGE gel to a support such as nitrocellulose membranes or Immobilon sheets, involves the application of a current that is perpendicular to the gel. This causes the proteins to travel out of the gel and onto the solid support, where they become covalently bound. This procedure is potentially very powerful if used in concert with immunological detection methods, which are described in Chapter 9.

Ideally, the blotting procedure should result in the transfer of all proteins regardless of their physicochemical properties. Unfortunately, this is generally not the case, as it is difficult to find a single set of blotting conditions that suit both the removal of the proteins from the gel as well as their binding to the solid support. Consequently, it is not wise to assume that the correct and complete transfer of all proteins present in the SDS–PAGE gel has necessarily been achieved. Many types of blotting apparatus are commercially available for the electrophoretic transfer of proteins. The exact blotting procedure will depend on the specific apparatus used and should be provided by the manufacturers.

An alternative method for the recovery of proteins from electrophoretic gels is electroelution.[38,39] This technique is similar to Western blotting, except that the proteins are eluted into a buffer contained in a series of elution chambers. The purified protein can then be analysed by a wide range of techniques, including mass spectrometry, amino acid analysis and peptide mapping.[40]

To our knowledge, these types of procedures have not yet been widely adopted in the field of Maillard chemistry, despite their enormous potential value.

7 Getting Started – in a Nutshell

When embarking upon an electrophoretic separation of Maillard reacted proteins, the first decision to be made is how best to resolve the protein mixture – will the optimal resolution be achieved by separation on the basis of size, charge or both size and charge? Having decided this, it will be possible to select the type of gel to be used. Separation by size can be done by SDS–PAGE gel and separation by charge by IEF. If both separation modes are to be used, then native gel electrophoresis will allow the retention of the biological activity of the sample, and 2-DE will enable the separation of complex mixtures. Of course, there are a number of variations on these methods; information regarding these less routine methodologies can be obtained from any electrophoresis text (see Further Reading).

The next stage of method development involves the selection of a gel pore size. For the separation of proteins on the basis of size, the pore size, that is the acrylamide and bisacrylamide content of the gel, should be adjusted to the size of the proteins to be analysed. For the separation of proteins that are of vastly different sizes, gradient gels should provide the best resolution without losing the smallest proteins off the bottom of the gel or leaving the largest at the top of the gel. Instead of consisting of a uniform pore size, the pore size of gradient gels gradually decreases as the proteins move through the gel. These gels are available commercially and provide an accurate picture of the nature of the sample in a single electrophoretic run.

Electrophoretic gels can be poured in the research laboraory, either with a uniform concentration of acrylamide or as a gradient gel. However, prepoured gels, for example those provided by Gradipore or BioRad, reduce the labour required, avoid exposure to toxic compounds, such as acrylamide monomer, and provide reliable and reproducible results. At the time of writing, prepoured minigels (approximate dimensions: 80 mm × 60 mm × 1 mm) have proven to be the most economical means of electrophoretic analysis. Unfortunately, the authors have not yet found it economically feasible to purchase large prepoured gels for routine use in the research laboratory, although, as the technology for pouring gels improves, the price will no doubt become more competitive. In some instances, minigels do not provide adequate separation of high molecular weight proteins. In such instances, large 'home-made' gels (approximate dimensions: 160 mm × 180 mm × 1.5 mm) have provided optimum resolution (see Case Study 7.1).

For the separation of protein mixtures on the basis of their charge, isoelectric focusing, native gel electrophoresis or, in combination with SDS–PAGE, 2-DE can be used. The pore size of IEF gels is generally selected such that all of the proteins to be analysed can move freely throughout the gel. In other words, the mobility, and therefore separation, of a protein mixture is limited only by the charge of the individual proteins and not by their molecular size. In order to create a pH gradient which will provide adequate separation, both for use on its own or as part of a 2-DE method, the appropriate pH range for the carrier ampholytes must be selected to match the sample composition. For example, if the approximate pI values of your reaction mixture are unknown, a wide range pH gradient (*e.g.* pH 3–10) may be a good starting point. If the sample mixture appears across a small range of the gel, for example between pH 4–5, then a second electrophoretic gel using a narrow pH range (*e.g.* pH 4–7) will provide better resolution.

One of the great advantages of electrophoresis as an analytical tool is that the equipment required is relatively inexpensive, and within the reach of most research laboratories. Unlike much of the instrumentation described in this book, protein electrophoresis equipment can be purchased readily. This, coupled with the ability to analyse intact proteins in a variety of different analytical modes, suggests a productive future for electrophoretic analysis in the analysis of food proteins.

8 Further Reading

B.D. Hames (ed.), *Gel Electrophoresis of Proteins*, 3rd edition, Oxford University Press, New York, 1998.

M.J. Dunn, *Gel Electrophoresis: Proteins*, BIOS Scientific, Oxford, 1993.

T. Rabilloud (ed.), *Proteome Research: Two-Dimensional Gel Electrophoresis and Identification Methods*, Springer, Heidelberg, 2000.

A. Gorg, C. Obermaier, G. Boguth, A. Harder, B. Scheibe, R. Wildgruber and W. Weiss, 'The Current State of Two-dimensional Electrophoresis with Immobilized pH Gradients', *Electrophoresis*, 2000, **21**, 1037.

9 References

1. M. Guzman-Gonzalez, F. Morais, M. Ramos and L. Amigo, *J. Sci. Food Agric.*, 1999, **79**, 1117.
2. M. Gil and C. Sarraga, *Food Biotechnol. (NY)*, 1997, **11**, 59.
3. N.F.S. Gault and R.A. Lawrie, *Meat Sci.*, 1980, **4**, 167.
4. W. Weiss, G. Huber, K.-H. Engel, A. Pethran, M.J. Dunn, A.A. Gooley and A. Goerg, *Electrophoresis*, 1997, **18**, 826.
5. E. Johansson, G. Svensson and W.K. Heneen, *Acta Agric. Scand., Sect. B*, 1995, **45**, 112.
6. P.L. Weegels, R.J. Hamer and J.D. Schofield, *J. Cereal Sci.*, 1996, **23**, 1.
7. G.L.W. Simpson and B.J. Ortwerth, *Biochim. Biophys. Acta*, 2000, **1501**, 12.
8. N. Sakata, Y. Sasatomi, S. Ando, J. Meng, Y. Imanaga, N. Uesugi and S. Takebayashi, *Connect. Tissue Res.*, 2000, **41**, 117.
9. K.W. Lee, C. Simpson and B. Ortwerth, *Biochim. Biophys. Acta*, 1999, **1453**, 141.
10. E.B. Frye, T.P. Degenhardt, S.R. Thorpe and J.W. Baynes, *J. Biol. Chem.*, 1998, **273**, 18714.
11. K. Nakamura, Y. Nakazawa and K. Ienaga, *Biochem. Biophys. Res. Commun.*, 1997, **232**, 227.
12. S.E. Fayle, J.A. Gerrard, L. Simmons, S.J. Meade, E.A. Reid and A.C. Johnston, *Food Chem.*, 2000, **70**, 193.
13. L. Pellegrino, M. van Boekel, H. Gruppen, P. Resmini and M.A. Pagani, *Int. Dairy J.*, 1999, **9**, 255.
14. S. Hill and A.M. Easa in *The Maillard Reaction in Foods and Medicine*, eds. J. O'Brien, H.E. Nursten, M.J.C. Crabbe and J.M. Ames, Royal Society of Chemistry, Cambridge, 1998, 133.
15. M. Friedman, *J. Agric. Food Chem.*, 1996, **44**, 631.
16. K. Ando, M. Beppu, K. Kikugawa, R. Nagai and S. Horiuchi, *Biochem. Biophys. Res. Commun.*, 1999, **258**, 123.
17. K.-W. Lee, V. Mossine and B.J. Ortwerth, *Exp. Eye Res.*, 1998, **67**, 95.
18. M.L. Plater, D. Goode and M.J.C. Crabbe, *Ophthalmic Res.*, 1997, **29**, 421.
19. H.K. Pokharna and L.A. Pottenger, *J. Surg. Res.*, 2000, **94**, 35.
20. Y.-T. Ho, S. Ishizaki and M. Tanaka, *Food Chem.*, 2000, **68**, 449.
21. R. Groubet, J.M. Chobert and T. Haertle, *Sciences des Aliments*, 1999, **19**, 423.
22. S.E. Fayle, J.P. Healy, P.K. Brown, E.A. Reid, J.A. Gerrard and J.M. Ames, *Electrophoresis*, 2001, **22**, 1518.

23. B.D. Hames, *Gel Electrophoresis of Proteins*, 3rd edition, Oxford University Press, New York, 1998.
24. Q. Shi and G. Jackowski in *Gel Electrophoresis of Proteins: A Practical Approach*, ed. B. D. Hames, Oxford University Press, New York, 1998, 1.
25. M. Tanaka, *Food Ingredients J.*, 2000, **185**, 23.
26. A. Handa and N. Kuroda, *J. Agric. Food Chem.*, 1999, **47**, 1845.
27. H. Saeki, *Nahrung*, 1998, **42**, 240.
28. A. Kato, H.R. Ibrahim, S. Nakamura and K. Kobayashi, *Egg Uses Process. Technol.*, 1994, 250.
29. S.E. Hill, J.R. Mitchell and H.J. Armstrong in *Gums and Stabilisers in the Food Industry*, eds. G.O. Phillips, D.J. Wedlock and P.A. Williams, Elsevier, London, 1992, 471.
30. P. Walstra in *Food Chemistry*, ed. O. R. Fennema, Marcel Dekker, New York, 1996, 95.
31. Y. Kato, K. Watanabe and Y. Sato, *Agric. Biol. Chem.*, 1983, **47**, 1925.
32. T. Rabilloud, *Anal. Chem.*, 2000, **72**, 48A.
33. T. Rabilloud, *Electrophoresis*, 1990, **11**, 785.
34. S. Fazekas de St Groth, R.G. Webster and A. Datyner, *Biochim. Biophys. Acta*, 1963, **71**, 377.
35. W.F. Patton, *Electrophoresis*, 2000, **21**, 1123.
36. G. Munch, Y. Taneli, E. Scharaven, U. Schindler, R. Schinzel, D. Palm and P. Riederer, *J. Neural. Transm.*, 1994, **8**, 193.
37. M.J. Dunn, *Gel Electrophoresis: Proteins*, BIOS Scientific Publishers Ltd, Oxford, 1993.
38. M. Shoji, M. Kato and S. Hashizume, *J. Chromatogr., A*, 1995, **698**, 145.
39. H.-T. Chang, A.L. Yergey and A. Chrambach, *Electrophoresis*, 2001, **22**, 394.
40. R.J. Simpson, R.L. Moritz and L.D. Ward, *Anal. Sci.*, 1991, **7**, 933.

CHAPTER 8

Capillary Electrophoresis

1 Introduction

As was discussed in Chapter 7, the electrophoretic separation of protein mixtures has traditionally been performed using polyacrylamide gels. An alternative to these methods is capillary electrophoresis (CE). Unlike the slab gel methodology, described in Chapter 7, CE is ideally suited to the analysis of small molecules such as the non-volatile colour compounds produced by the Maillard reaction.[1] Although CE can be used for the analysis of large macromolecules, in the Maillard arena its use has predominantly been limited to the analysis of low molecular weight compounds.[2-6]

CE offers a number of advantages over slab gel electrophoresis since for most samples CE analysis times are extremely short, yet selectivity and sensitivity remain high. In addition, the amount of sample required is small, generally much less than 1 mg, and automated instrumentation allows for multiple analyses providing quantifiable results.

2 How Does CE Work?

CE separations occur in small diameter (<100 μm) fused-silica capillaries, similar to those used for GC, although much shorter, typically less than 100 cm in length (Figure 8.1).

As for GC and HPLC, some CE separations are achieved by the partitioning of the sample components between a mobile phase and a stationary phase. The mobile phase in CE separations is usually a buffer solution, which may or may not contain any of a wide range of additives used to modify the buffer's characteristics.

Whereas HPLC uses an external pump to push the mobile phase through the column, the movement of the mobile phase in CE is controlled by a phenomenon known as the electro-osmotic flow (EOF), a unique feature of CE. The inner surface of the capillary is negatively charged and therefore attracts positively charged counter-ions from the buffer solution, which accumulate immediately adjacent to the wall of the capillary, as shown in Figure 8.2. When a voltage is applied, the positively charged counter-ions are pulled through the capillary by

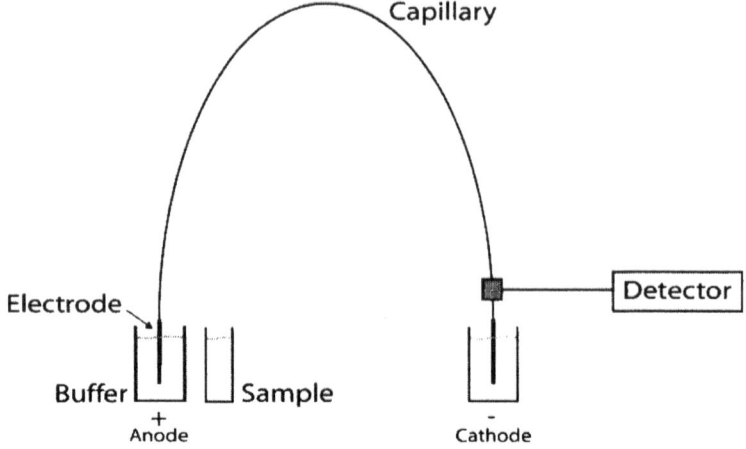

Figure 8.1   *Diagram of a typical CE instrument*

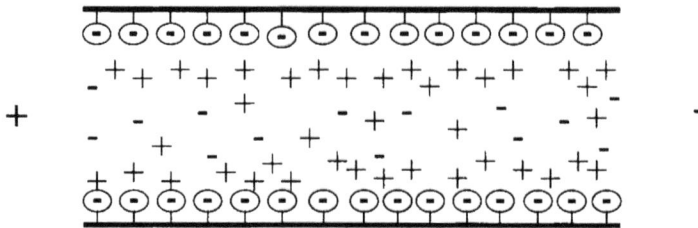

Figure 8.2   *Profile of a fused-silica capillary showing the negatively charged inner wall and the arrangement of the buffer ions within the capillary – the positively charged ions accumulate immediately adjacent to the wall of the capillary*

their attraction to the negatively charged cathode. As the positive counter-ions are surrounded by the buffer containing the sample, their movement drags the bulk solution with them, toward the cathode, and through the detector. As the sample is pulled through the capillary by the EOF, a flat flow profile is achieved. The resulting peaks are much sharper than those produced by HPLC, in which the external pump system pushes the sample through the column, resulting in a parabolic flow shape, as friction at the wall of the column limits the forward movement of the sample. A comparison of the flow profiles and resulting peak shapes of CE and HPLC are shown in Figure 8.3.

As the speed of the analysis can impact upon the resolution of the separation, it is desirable that the analyst has some control over the speed of the EOF. This can be achieved by manipulating a number of parameters, the most obvious of which is pH. The silanol groups on the inner wall of the fused-silica capillaries are negatively charged above ~pH 4. If a buffer with a pH < 4 is used as the mobile phase, the silanol groups will be protonated and will no longer be charged. This will result in an EOF that is slow, increasing the time that the sample spends within the capillary and, theoretically, improve the separation.

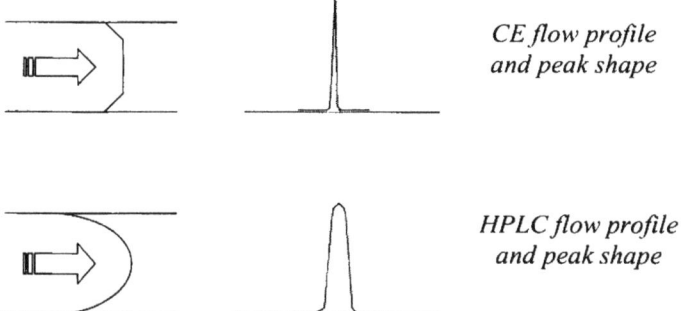

*CE flow profile
and peak shape*

*HPLC flow profile
and peak shape*

Figure 8.3 *Flat flow profile of CE, and resulting sharp peak shape, compared with the parabolic flow shape and broader peak resulting from HPLC*

Other methods of manipulating the EOF include: varying the ionic strength of the buffer, as buffers containing a high number of ions mask the negative charge of the capillary wall; altering the capillary temperature or the applied voltage; or the use of coated capillaries or buffer additives. These parameters will be discussed further in the method development section of this chapter.

3 Separation Techniques

Optimising the separation of molecules by CE requires altering the partitioning of sample components between the mobile phase and the stationary phase. This is generally achieved by the addition of various buffer-modifying agents. Alternatively, the components may simply be separated according to their charge. This method of separation, known as capillary zone electrophoresis, will be discussed first.

Capillary Zone Electrophoresis (CZE)

Capillary zone electrophoresis (CZE), sometimes known as free solution capillary electrophoresis (FSCE), is the most widely used of all of the CE techniques. As in traditional electrophoresis, separation of the sample is achieved by differences in charge. As previously described, the EOF is produced by the attraction of positively charged counter-ions toward the negatively charged electrode and is responsible for dragging the sample mixture through the capillary toward the detector. If the sample contains positively charged species, these will experience an additional attraction toward the cathode that will result in a greater mobility – they will travel toward the cathode at a faster rate than that of the EOF. Negatively charged molecules will be repulsed by the negatively charged cathode, so travel at a slower rate than the EOF and pass through the detector after any positively charged molecules. Neutral molecules will have no additional attraction or repulsion, so travel at the same speed as the EOF; this means that they reach

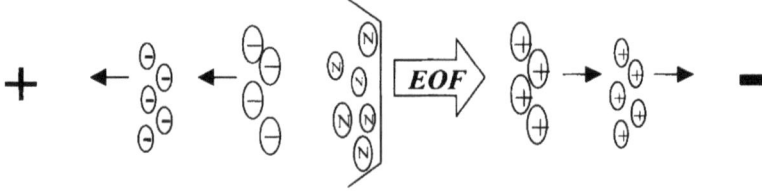

Figure 8.4   *CZE separates molecules according to their charge – positively charged molecules travel the fastest, ahead of the EOF, neutrals at the same speed as the EOF and negatively charged molecules travel at a slower rate than the EOF*

the detector after any positively charged molecules, but before any negatively charged species. This is shown schematically in Figure 8.4.

Whereas slab gel electrophoresis enables the analysis of only positive or negative ions at any one time, CE is capable of separating both positive and negative ions, as well as neutral molecules, in a single run. This is due to the EOF, which causes the movement of nearly all species, regardless of charge, toward the detector. Although neutral molecules can be separated from charged ions, they cannot be separated from each other as there is no difference in charge. In order to separate these molecules by CZE, they must first be derivatised in a way that produces a charged ion.

Although CZE has been used for the analysis of a vast array of compounds in foods, ranging from amino acids[7] and peptides[8] to free fatty acids,[9] minerals[10] and proteins,[11] few studies have utilised this potentially powerful technique in the study of the Maillard reaction.[1,2,6,12–15]

CASE STUDY 8.1 – HPLC *versus* CZE?

A brief literature search revealed that very few research laboratories have employed CE in their studies of Maillard reaction systems. Of those that have, the majority of these studies have utilised CZE as the mode of analysis. For example, Royle *et al.*[6] have compared the abilities of CZE and reverse-phase HPLC to separate Maillard reaction products arising from two model systems, containing either glycine–glucose or glycine–xylose. CZE was found to be a much faster and more efficient method for the separation of Maillard reaction products, with many more products being resolved than by reverse phase HPLC. Further separation of the reaction products of the more reactive glycine–xylose system was achieved by ultrafiltration, separating them into three molecular weight ranges (< 1000 Da, 1000–3000 Da and > 3000 Da). The two higher molecular weight fractions, consisting of melanoidins, were found to migrate as broad humps, as shown in Figure 8.5, whereas the lower

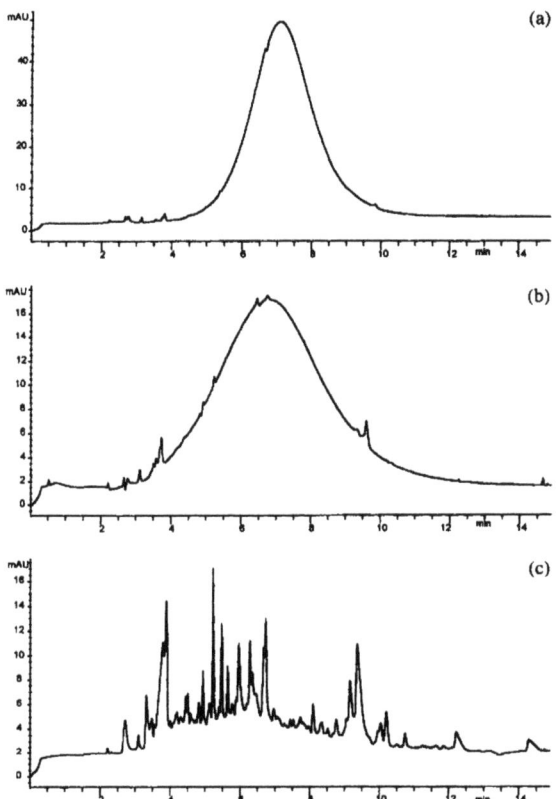

Figure 8.5   *E-grams of ultrafiltration fractions from a xylose–glycine model system after 2 hours under reflux at pH 5, with UV detection at 200 nm, (a) above 3000, (b) between 1000 and 3000 and (c) below 1000 Daltons molecular weight*
(Reproduced with permission[6])

molecular weight fraction resulted in an electropherogram (e-gram) that contained many more discernible peaks with almost baseline resolution.

Likewise, an examination of an extrusion-cooked starch–lysine–glucose model system by CZE indicated that this technique gave separations superior to those achieved by either reverse phase HPLC or ion exchange HPLC.[1] It was noted however that, unlike HPLC, the CE technique cannot be scaled up to give preparative separations.

Earlier studies have also compared the merits of CE with HPLC.[2,15] In these investigations, spray-dried glucose–glycine and refluxed 5-hydroxymethyl-furfural (HMF)–glycine model systems were studied. CE was again found to perform better than HPLC with regard to both resolution and analysis time. Tomlinson and co-workers[12] have employed CE, with online detection by electrospray mass spectrometry, in order to analyse the reaction of ^{15}N-labelled glycine and HMF.

Micellar Electrokinetic Chromatography (MEKC)

Micellar electrokinetic chromatography (MEKC), also known as micellar electro-kinetic capillary chromatography (MECC), is one of the many methods used to separate reaction mixtures by partitioning between a mobile phase and a stationary phase. This methodology requires the addition of a surfactant, a compound which contains both a water-soluble (hydrophilic) and a fat-soluble (hydrophobic) region, such as SDS, to the running buffer. As the name suggests, the surfactant must therefore be added so that its concentration results in the formation of micelles – it must be present in a concentration above the 'critical micellar concentration' (cmc), which for SDS is 8 mM. Above this concentration, clusters of the surfactant molecules gather together to form an aggregate, known as a micelle, in which the fat-soluble end of each molecule is oriented toward the middle of the micelle and is protected from the aqueous buffer. The micelle is then able to serve as a pseudo-stationary phase for the separation, travelling either faster or slower than the EOF depending on the charge of the surfactant used (see Figure 8.6). If the surfactant is SDS, which is negatively charged, the micelle will be repelled by the negatively charged cathode, and so will travel toward the detector at a slower rate than that of the EOF.

MEKC separation is, therefore, based on the degree of hydrophobicity, since hydrophobic molecules will interact with the micelle whereas the hydrophilic molecules will remain in the buffer and travel at the speed of the EOF. As well as

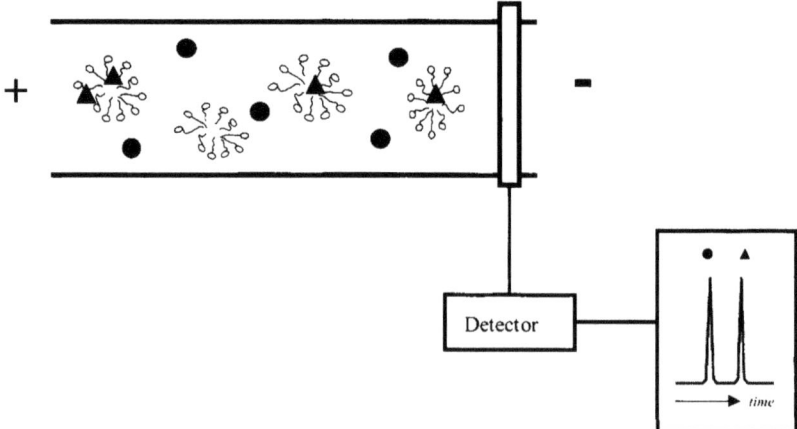

Figure 8.6   *In MEKC, sample components are separated by partitioning between the buffer and a micelle. For example, hydrophilic molecules (•) remain in the buffer and are eluted first, whereas hydrophobic molecules (▲) elute last, since they associate with the hydrophobic interior of the micelle and are temporarily retained by the capillary*

separating molecules by their differences in hydrophobicity, separation by MEKC can be based on differences in ionic attraction or hydrogen bonding, depending on which additive is used.

This form of separation finds particular use in the separation of neutral molecules, since the separation is not based on charge. MEKC has also been used for the separation of proteins,[16] synthetic colours,[17] artificial sweeteners,[18] flavanoids,[19] catechins[20] and phenolics.[21] This methodology has not yet been adopted by investigators in the Maillard field.

Capillary Gel Electrophoresis (CGE)

Capillary gel electrophoresis (CGE) is a technique which is comparable with the standard SDS–polyacrylamide gel electrophoresis (SDS–PAGE) technique described in Chapter 7. In both CGE and SDS–PAGE, sample components are separated by differences in size. Unlike the standard slab gel format, where separation occurs on a solid gel support, CGE generally involves the use of polymer networks, which are not the solid gels that the name suggests. Rather, they are polymer solutions that are just fluid enough to be capable of flowing through the capillary, although some polymers can be crosslinked within the capillary and are not designed to be removed. The resulting polymer network is ideal for the separation of large macromolecules, such as proteins, since it behaves as a molecular sieve. Small proteins are able to pass through the capillary relatively unhindered, whereas large proteins take much longer to reach the detector, as shown in Figure 8.7.

The various polymers that have been used in CGE include crosslinked or linear polyacrylamide, dextran, polyethylene oxide, agarose and methylcellulose

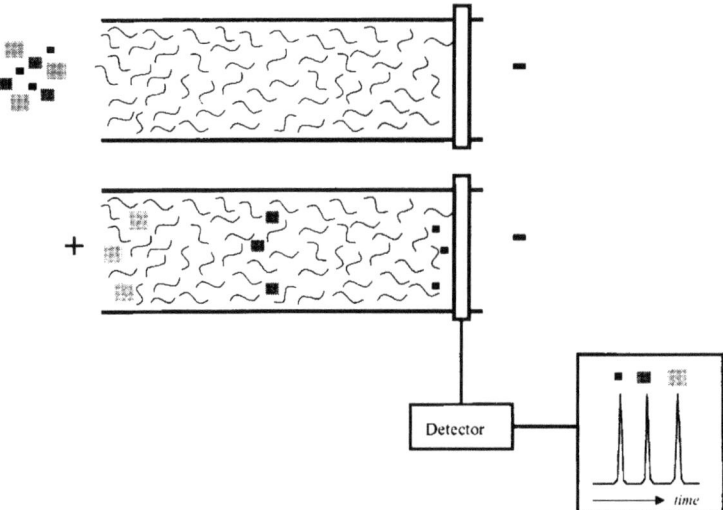

Figure 8.7 *CGE separates proteins and peptides on the basis of size by passing the sample components through sieve-like polymer networks*

derivatives, such as hydroxymethylcellulose. For standard protein separations it is possible to purchase a preprepared kit, which obviates the need to determine which additive, and at what concentration, is best suited to a particular protein mixture (see Figure 8.8).

CGE offers numerous advantages over the traditional slab gel method: the sample requirement is small; it uses a fraction of the amount of consumables required for slab gel electrophoresis; and, coupled with on-line detection, it allows rapid, quantitative analysis. Two disadvantages of CGE, when compared with the slab gel method traditionally used, are the difficulties of collecting the purified analytes and scaling up to a preparative scale.

CGE is generally used for the analysis of proteins and peptides,[22] although very few examples of the application of CGE to the analysis of food proteins are present in the literature.

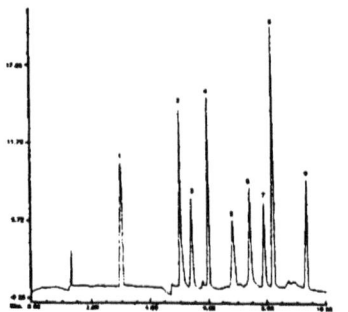

Figure 8.8   *E-gram showing the separation of protein standards using the BioRad CE–SDS Protein Kit*
(Reproduced with permission from BioRad)

Capillary Isoelectric Focusing (CIEF)

Capillary isoelectric focusing (CIEF) is analogous to the traditional IEF technique, described in Chapter 7, where proteins and peptides are separated on the basis of their overall charge. More accurately, separation is due to differences in isoelectric point (pI), which is the pH at which the protein has an overall charge of zero (see Figure 7.5).

In order to separate a mixture of proteins by their pI, two requirements must be met – removal of the EOF by coating the inner wall of the capillary, hence masking the negative charge of the silanol groups, and the preparation of a pH gradient within the capillary. The pH gradient is prepared by filling the capillary with a solution containing both the sample and a range of ampholytes, compounds that contain both positive and negative moieties, which span the desired pH range. Having positioned an acidic solution at the anode, and a basic solution at the cathode, the application of a current will result in the migration of the ampholytes such that a pH gradient is formed. The sample components will also migrate.

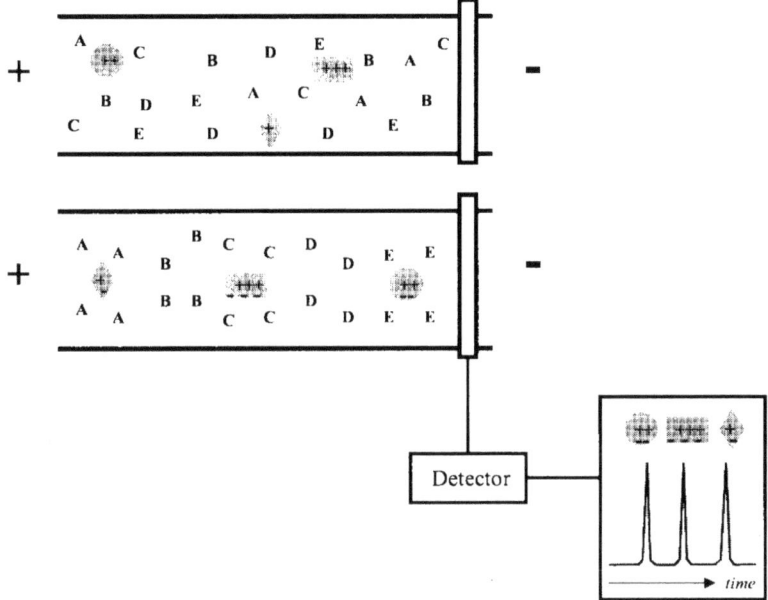

Figure 8.9 *The capillary is filled with a solution containing both the sample and the ampholytes (labelled A–E). The application of a current results in the formation of a pH gradient and the migration of the sample components to a point in the gradient where the pH equals its pI*

They will travel along the pH gradient until they reach the pH that equals their pI, as can be seen in Figure 8.9. As each protein with a different pI will migrate to a different position in the pH gradient, stopping when they have an overall charge of zero, the proteins will ultimately be positioned at various points along the capillary.

Having separated the proteins, and since it is not possible to view the whole length of the capillary, it is now necessary to push the resolved proteins through the detector so that their position in the pH gradient can be monitored. This is often done by the application of pressure to one end of the capillary.

Although CIEF is commonly used in the medical sciences,[23–26] it is yet to be adopted in the Maillard field. Since it has been suggested that the pI of a protein changes when it undergoes Maillard chemistry,[23,27] this methodology could prove to be a powerful tool for the analysis of protein-derived Maillard reaction products.

CASE STUDY 8.2 – The Analysis of Furosine

The compound ε-*N*-2-furoylmethyl-L-lysine, commonly known as furosine, is the acid-hydrolysed product of the Amadori compound derived from reaction of a lysine residue. Furosine, shown in Figure 8.10, has attracted a great deal of attention in the literature as it provides a useful marker for evaluating the

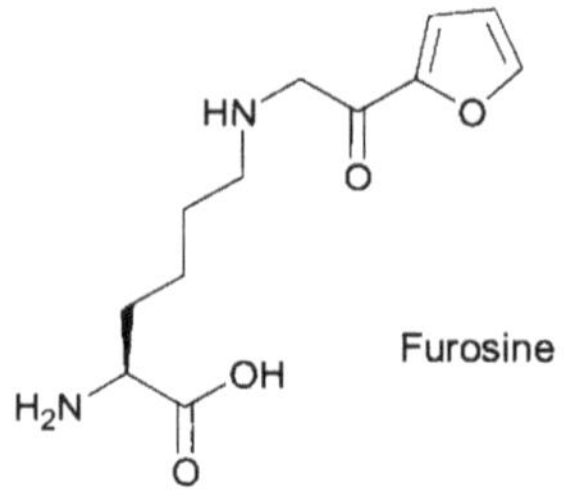

Figure 8.10   *Structure of furosine (ε-N-2-furoylmethyl-L-lysine)*

quality of many foods.[27-31] The presence of furosine has traditionally been determined by HPLC.[32]

A CE method for the determination of furosine was published by Tirelli in 1998.[13] This method was used for the analysis of a range of heat-treated foods, including various cheeses, milk and durum wheat products. It compared favourably with the validated HPLC methodology traditionally used and was found to have some advantages over the HPLC technique, namely the speed of analysis and the low cost involved in CE analyses.

4   Detection

As CE analyses are performed on very small sample sizes, a very sensitive detection method is crucial to the success of the experiment. The most common form of detection used in CE is UV detection and the resulting trace is known as an electropherogram, or e-gram. Examples of typical e-grams can be seen in Case Studies 8.1 and 8.3. Other detectors used with CE include any of the common HPLC detectors, such as those using fluorescence, described in Chapter 5, or mass spectrometry, discussed in Chapter 6.

5   Protein Analysis

CE is suitable for the analysis of both low molecular weight compounds and large macromolecules, such as proteins. Although CE offers numerous advantages over traditional methods of analysing complex mixtures, such as those resulting from the Maillard reaction, it also suffers from a number of disadvantages that merit some discussion. The most relevant disadvantage when studying protein mixtures is the potential for the adsorption of proteins onto the capillary wall.

Adsorption of proteins to the inner wall of the capillary is undesirable since it can result in broad sample peaks, peak tailing, non-quantitative analyses and reduced separation efficiency. The primary causes of adsorption include hydrophobic interactions and electrostatic interactions between the positively charged sample and the negatively charged capillary wall. For small molecules, these interactions can be minimised by working at a pH where the charge of the molecule is completely eliminated. Interactions between the capillary wall and larger molecules, such as proteins and large peptides, however, are not so easily removed, as the protein comprises many amino acids that often have charged or hydrophobic side chains. Simply altering the pH may reduce the number of charges present on the molecule as a whole, but will never entirely eliminate them. These localised charges and areas of hydrophobicity are available to interact with the inner wall of the capillary, causing the protein to stray from its path toward the detector. Thus, in order to apply CE to the study of protein mixtures, the adsorption of proteins to the capillary wall must be minimised. Various approaches have been proposed for the prevention of these interactions, most of which involve the masking of the negative charge carried by the wall of the capillary. Common methods include using high concentration buffers, buffers of extreme pH, and either temporary (with the use of buffer additives) or permanent capillary coatings. Case study 8.3 illustrates the use of a temporary (or dynamic) coating using the polymeric additive hydroxymethyl cellulose.

CASE STUDY 8.3 – Protein-derived Maillard Reaction Products

Very few CE studies of Maillard reaction systems containing intact proteins have appeared in the literature, and only a handful of these have stemmed from the field of food science – these have generally involved the study of milk proteins. Jones *et al.*[33] analysed the reaction of lactose with the protein β-lactoglobulin, both present in milk. Using a citrate buffer containing the buffer additive hydroxymethyl cellulose they were able to show the presence of up to three Maillard reaction products, appearing as secondary peaks migrating after β-lactoglobulin. The identities of these peaks were confirmed as the consecutive formation of lactuloselysine adducts by electrospray mass spectrometry. As can be seen in Figure 8.11, the secondary peaks were present in the skimmed milk powder, but not in fresh skimmed milk. Thus, it is likely that the manufacturing process of skimmed milk powder, typically involving spray drying, encourages Maillard chemistry (see also Case Study 6.2).

Similar results were obtained by De Block *et al.*[34] In this instance, although CE and slab gel isoelectric focusing (IEF) were able to demonstrate the presence of Maillard reaction products, these products could not be detected by HPLC, fast polymer liquid chromatography or SDS–PAGE. Hence this study demonstrates the value of both CE and IEF when studying Maillard reaction systems.

A more recent study has been performed in the authors' laboratories, investigating the reaction of a model protein, ribonuclease A (RNAse A), with a range of carbonyl-containing compounds that are commonly found in foods.[35] In each case, product peaks appeared on the trailing edge of the RNAse A peak. An example is shown in Figure 8.12. The combination of CZE, SDS–PAGE and 2-DE demonstrated that the formation of multimeric protein was increasing with time, but that a substantial amount of monomeric

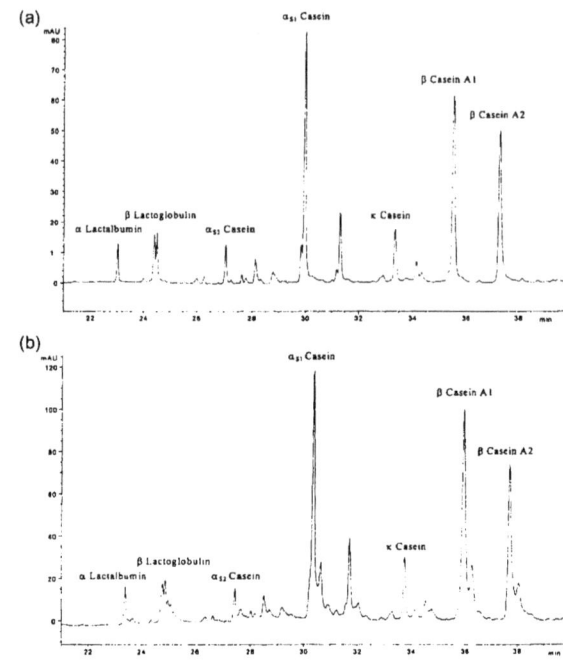

Figure 8.11   *E-grams of (a) fresh skimmed milk and (b) skimmed milk powder*
(Reproduced with permission[33])

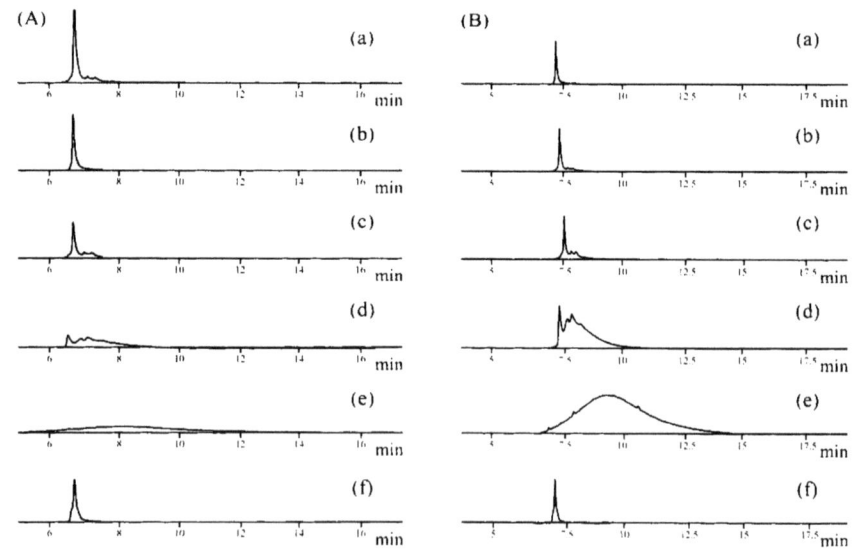

Figure 8.12   *E-grams of a DHA–RNAse A model system, incubated at 25 °C for up to 42 days; (A) each of the e-grams shown in same scale, (B) each e-gram is shown in full scale. Time of reaction, (a) day 1, (b) day 3, (c) day 7, (d) day 14, (e) day 42 and (f) protein-only standard, incubated for 42 days*
(Reproduced with permission[35])

protein was also present. Products with a range of isoelectric points, as shown by both CE and 2-DE, were also produced. It was hypothesised that the product peaks in the CE e-gram were due to the formation of compounds that do not contribute to protein crosslinking, such as carboxymethyllysine.

CE was also found to provide a useful means of monitoring the rate of the Maillard reaction, such that the reaction rate at different temperatures, in the presence of various carbonyl-containing compounds, could be compared, as shown in Figure 8.13.

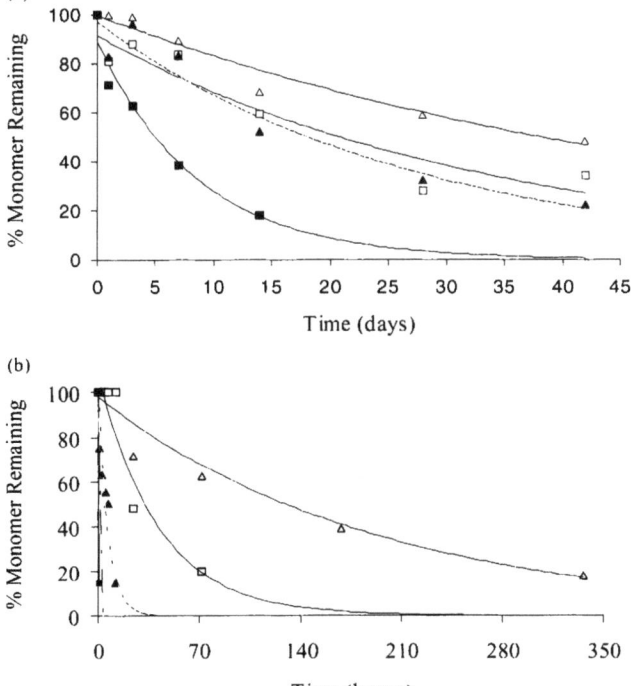

Figure 8.13 *Monomeric RNAse A content of (a) Glucose–RNAse A (△), Xylose– RNAse A (□), Ascorbic acid–RNAse A (▲) and DHA–RNAse A (■), incubated at 25 °C, with a line of best fit; (b) DHA–RNAse A samples incubated at 25 °C (△), 37 °C (□), 50 °C (▲) and 75 °C (■), with a line of best fit.*
(Reproduced with permission[35])

6   Getting Started – in a Nutshell

A useful first step for developing a CE method is to complete a review of the literature, as this will often provide starting parameters for most compounds. Unlike other techniques, CE analysis times, along with capillary conditioning times, are very short. It is therefore possible to test various modifications of the starting parameters and optimise your methodology in a very short space of time. A number of factors must be considered when developing a CE method. Firstly, the basis of separation must be decided – such as charge, hydrophobicity or pI – so that the specific CE methodology – CZE, MEKC, CGE or CIEF – can be selected. Secondly, the selection of running buffer is extremely important to the success of any CE separation. The buffer must have good buffering capacity at the chosen pH. If UV detection is to be used, the buffer must have low UV absorption at the detection wavelength. The buffer should also provide a stable environment for the analyte. Commonly used buffers include phosphate, citrate, Tris and borate buffers.

When working with large analytes, such as proteins, it is also worth considering the pI of the protein – by working at a pH which is below the pI of the protein, for instance, the net charge of the protein will be positive, causing it to migrate ahead of the EOF, thus shortening the analysis time.

Other factors which must be considered are the capillary temperature, sample concentration, injection method and volume, the voltage applied to the capillary and, if analysing proteins, a method of minimising protein adsorption to the capillary wall. An example of the effect which buffer pH or the applied voltage can have on the peak shape and the analysis time can be seen in Figure 8.14. For this particular analyte, the protein RNAse A, a pH of 2.5 with an applied voltage of 20 kV, was found to be optimal.

CE is a relatively new technique, and yet has already found considerable application in the Maillard field. It is likely to gain in importance as more Maillard researchers acquire access to CE instrumentation that allows them to analyse small amounts of sample extremely rapidly.

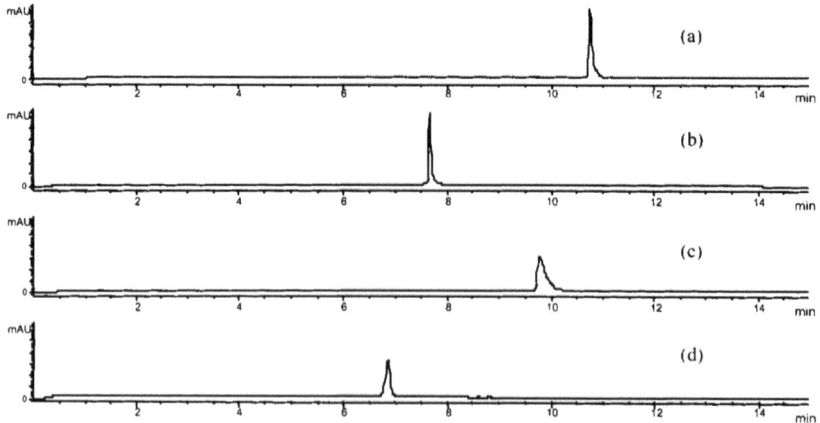

Figure 8.14   *The affect of pH and applied voltage on the migration time and peak shape. Conditions for each e-gram were (a) 15 kV at pH 2.5. (b) 20 kV at pH 2.5, (c) 15 kV at pH 2.0 and (d) 20 kV at pH 2.0*

7 Further Reading

R.A. Frazier, J.M. Ames and H.E. Nursten, *Capillary Electrophoresis for Food Analysis*, Royal Society of Chemistry, Cambridge, 2000.

H. Sorenson, S. Sorenson, C. Bjergegaard and S. Michaelson, *Chromatography and Capillary Electrophoresis in Food Analysis*, Royal Society of Chemistry, Cambridge, 1999.

J.P. Landers (ed.), *Handbook of Capillary Electrophoresis*, 2nd ed., CRC Press, Boca Raton, Florida, 1997.

Z. El Rassi and R.W. Giese, *Selectivity and Optimisation in Capillary Electrophoresis*, Elsevier, Oxford, 1997.

P. Camilleri (ed.), *Capillary Electrophoresis: Theory and Practice*, 2nd ed., CRC Press, Boac Raton, Florida, 1998.

8 References

1. J.M. Ames, A. Arnoldi, L. Bates and M. Negroni, *J. Agric. Food Chem.*, 1997, **45**, 1256.
2. A.J. Tomlinson, J.A. Mlotkiewicz and I.A.S. Lewis, *Food Chem.*, 1994, **49**, 219.
3. Z. Deyl, I. Miksik and R. Struzinsky, *J. Chromatogr.*, 1990, **516**, 287.
4. V.M. Hill, J.M. Ames, D.A. Ledward and L. Royle in *Maillard Reaction in Foods and Medicine*, eds. J. O'Brien, H.E. Nursten, M.J.C. Crabbe and J.M. Ames, Royal Society of Chemistry, Cambridge, 1998, 121.
5. A. Lapolla, D. Fedele, R. Aronica, O. Curcuruto, M. Hamdan, S. Catinella, R. Seraglia and P. Traldi, *J. Mass Spectrom.*, 1995, S69.
6. L. Royle, R.G. Bailey and J.M. Ames, *Food Chem.*, 1998, **62**, 425.
7. T. Tsuda and T. Hasumi, *Chromatography*, 1997, **18**, 39.
8. J.S. Madsen, T.O. Ahmt, J. Otte, T. Halkier and K.B. Qvist, *Int. Dairy J.*, 1997, **7**, 399.
9. C.W. Klampfl, W. Buchberger and P.R. Haddad, *J. Chromatogr. A*, 2000, **881**, 357.
10. Q. Yang, C. Hartmann, J. Smeyers-Verbeke and D.L. Massart, *J. Chromatogr. A*, 1995, **717**, 415.
11. Z. El Rassi and Y.S. Mechref in *Capillary Electrophoresis*, ed. P. Camilleri, CRC, Boca Raton, Florida, 1998, 273.
12. L.M. Benson, S. Naylor and A.J. Tomlinson, *Food Chem.*, 1998, **62**, 179.
13. A. Tirelli, *J. Food Protection*, 1998, **61**, 1400.
14. Z. Deyl, I. Miksik and J. Zicha, *J. Chromatogr.*, 1999, **836**, 161.
15. A.J. Tomlinson, J.P. Landers, I.A.S. Lewis and S. Naylor, *J. Chromatogr.*, 1993, **652**, 171.
16. J.-F. Fairise and P. Cayot, *J. Agric. Food Chem.*, 1998, **46**, 2628.
17. C.O. Thompson and V.C. Trenerry, *J. Chromatogr. A*, 1995, **704**, 195.
18. C.O. Thompson, V.C. Trenerry and B. Kemmery, *J. Chromatogr. A*, 1995, **694**, 507.
19. J.D. Fontana, M. Passos, M.H.R. Dos Santos, C.K. Fontana, B.H. Oliveira, L. Schause, R. Pontarolo, M.A. Barbirato, M.A. Ruggiero and F.M. Lancas, *Chromatographia*, 2000, **52**, 147.
20. B.C. Nelson, J.B. Thomas, S.A. Wise and J.J. Dalluge, *J. Microcolumn Sep.*, 1998, **10**, 671.
21. J. Summanen, H. Vuorela, R. Hiltunen, H. Siren and M.L. Riekkola, *J. Chromatogr. Sci.*, 1995, **33**, 704.
22. M. Cota-Rivas and B. Vallejo-Cordoba, *J. Capillary Electrophor.*, 1997, **4**, 195.
23. M. Sugano, H. Hidaka, K. Yamauchi, T. Nakabayashi, Y. Higuchi, K. Fujita, N. Okumura, Y. Ushiyama, M. Tozuka and T. Katsuyama, *Electrophoresis*, 2000, **21**, 3016.

24. A. Lupi, S. Viglio, M. Luisetti, M. Gorrini, P. Coni, G. Faa, G. Cetta and P. Iadarola, *Electrophoresis*, 2000, **21**, 3318.
25. S. Martinovic, L. Pasa-Tolic, C. Masselin, P.K. Jensen, C.L. Stone and R.D. Smith, *Electrophoresis*, 2000, **21**, 2368.
26. T. Manabe, H. Miyamoto, K. Inoue, M. Nakatsu and M. Arai, *Electrophoresis*, 1999, **20**, 3677.
27. N. Corzo, M. Villamiel, M. Arias, S. Jimenez-Perez and F.J. Morales, *Food Chem.*, 2000, **71**, 255.
28. M. Villamiel and N. Corzo, *Milchwissenschaft*, 2000, **55**, 90.
29. E. Guerra-Hernandez, N. Corzo and B. Garcia-Villanova, *J. Cereal Sci.*, 1999, **29**, 171.
30. C. Pompei and A. Spagnolello, *Meat Sci.*, 1997, **46**, 139.
31. S. Nardi, C. Calcagno, P. Zunin and F. Evangelisti, *Riv. Sci. Aliment.*, 1998, **27**, 29.
32. T. Henle, G. Zehetner and H. Klostermeyer, *Z. Lebensm. Unters. Forsch*, 1995, **200**, 235.
33. A.D. Jones, C.M. Tier and J.P.G. Wilkins, *J. Chromatogr. A*, 1998, **822**, 147.
34. J. De Block, M. Merchiers and R. Van Renterghem, *Int. Dairy J.*, 1998, **8**, 787.
35. S.E. Fayle, J.P. Healy, P.K. Brown, E.A. Reid, J.A. Gerrard and J.M. Ames, *Electrophoresis*, 2001, **22**, 1518.

CHAPTER 9

New Methodologies, New Approaches

1 Introduction

In Chapters 1 and 2, the complexities of the Maillard reaction and its many ramifications during food processing were introduced. The rest of the book has been devoted to a consideration of the many methods that can be applied to unravel the details of this intriguing chemical process. So far, the emphasis has been very much on how to separate the products of the reaction, employing a wide range of separation strategies. With the exception of the discussion of mass spectrometry in Chapter 6, less attention has been paid to the characterisation and structural elucidation of the reaction products.

Until recently, the difficulty in separating the complex mixtures of Maillard reaction products, derived either from model systems or *via* extraction from a foodstuff, has imposed a serious impediment to our understanding of the Maillard reaction. It is, therefore, not surprising that the bulk of the Maillard literature places a heavy emphasis on separation methodologies. In the last ten years, substantial progress has been made in the separation of individual reaction products. Many such products have been obtained in sufficient quantity for characterisation using the rigorous methods of chemical structure elucidation. In particular, nuclear magnetic resonance (NMR), a technique so widely used in other areas of chemistry, is beginning to unlock many of the mysteries of Maillard chemistry. Results from detailed NMR studies have begun to yield not only unambiguous structural information about previously unidentified reaction products, but also insights into the mechanisms by which these products may have been formed.

Techniques emerging from the biochemical sciences have also begun to find application in the Maillard field. In particular, immunochemistry – in which antibodies are generated that detect a particular product of interest by virtue of a very specific binding site – has begun to allow the detection of specific molecules of interest within a complex mixture. This elegant method obviates the need to separate Maillard reaction products from food, and has considerable potential in the monitoring of specific molecules of interest that may form during food processing.

A detailed discussion of the intricacies of these various new approaches of structure elucidation is beyond the scope of this text. However, the use of these methods in the Maillard field will be briefly reviewed, and the power of the approaches emphasised with a few state-of-the-art case studies. Those interested in the theory behind the techniques introduced in this chapter are referred‚to the cited literature and the introductory texts in the Further Reading section.

2 Structural Elucidation of Purified Maillard Products

Detailed structural elucidation of a molecule requires a purified sample, generally in milligram quantities. The precise criteria for purity and sample size are very much dependent on the specific technique to be employed, but obtaining even a milligram of a purified Maillard reaction product can prove extremely challenging. Advances in this field arise whenever a method is developed that allows the isolation of a molecule in large enough quantities for study, or when a method is developed with improved sensitivity, such that accurate information can be obtained on smaller quantities.

Most Maillard research has traditionally relied on the detection of reaction products by a characteristic absorbance of light, often in the UV or visible range. As we have seen, identification is often achieved by matching the absorption spectrum (and separation parameters, such as retention time on a specified HPLC column) of the compound of interest with an authentic sample. Using such methods, many molecules have been identified without ever having been isolated. It has, therefore, been difficult to employ the two main methods employed by chemists and biochemists to elucidate molecular structure: X-ray crystallography and NMR.

X-ray crystallography requires that the compound of interest exist in crystalline form, a condition not met by most Maillard reaction products. However, X-ray methods have been employed in occasional studies.[1] Other workers have used X-ray methods not to determine the structure of a Maillard reaction product, but to detect a change in ordering of molecules in a solid sample that is thought to be undergoing Maillard chemistry.[2-5]

Unlike X-ray crystallography, NMR spectrometry is applicable to any molecule that can be dissolved in an appropriate NMR-compatible solvent, and is finding increasing use in the Maillard field.[1,6-28] Most studies involve isolation of the compound of interest, followed by analysis by one-dimensional and/or two-dimensional NMR methods. The vast majority of workers to date have employed proton (^{1}H) or carbon (^{13}C) NMR of molecules in solution. An example of the level of structural information that can be obtained is given in Case study 9.1. There are also occasional studies of particular molecules which have examined the structure of Maillard products by phosphorus (^{31}P)[29] or fluorine (^{19}F)[30] NMR, or in the solid state.[31,32]

CASE STUDY 9.1 – Structural Elucidation of the Melanoidins

Much of the analysis of the Maillard reaction to date has focused on early reaction products, which tend to be of low molecular weight and are comparatively easy to identify. The later stages of the reaction, particularly the formation of the high molecular weight melanoidins, have received less attention, mainly due to the extreme difficulty in separating and analysing a complex, heterogeneous mixture of polymeric material. Tressl and co-workers[22] have made substantial progress on the structural elucidation of melanoidins by carrying out model experiments in which the polycondensation of variously substituted pyrroles were examined in combination. The structures of the resulting mixtures were analysed by mass spectrometry and NMR. ^{13}C-labelling experiments were also employed to characterise the products further. The results of these experiments suggested that regular oligomers containing up to 15–30 methine-bridged-pyrroles were formed (Figure 9.1). These results were used to propose a structure for native melanoidins (Figure 9.2).

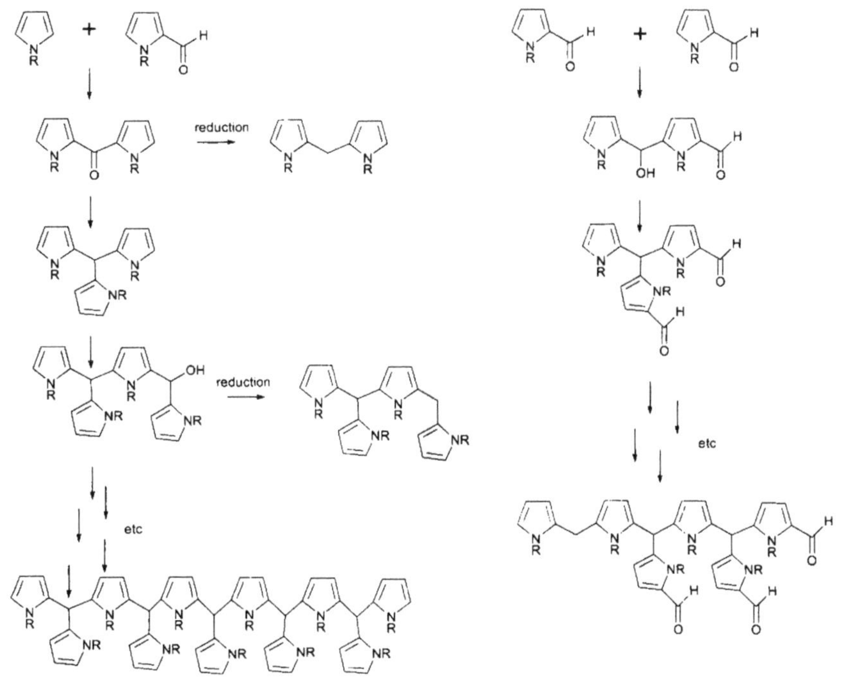

Figure 9.1 *Proposed mechanism of formation of melanoidin-like Maillard polymers from pyrroles[22]*

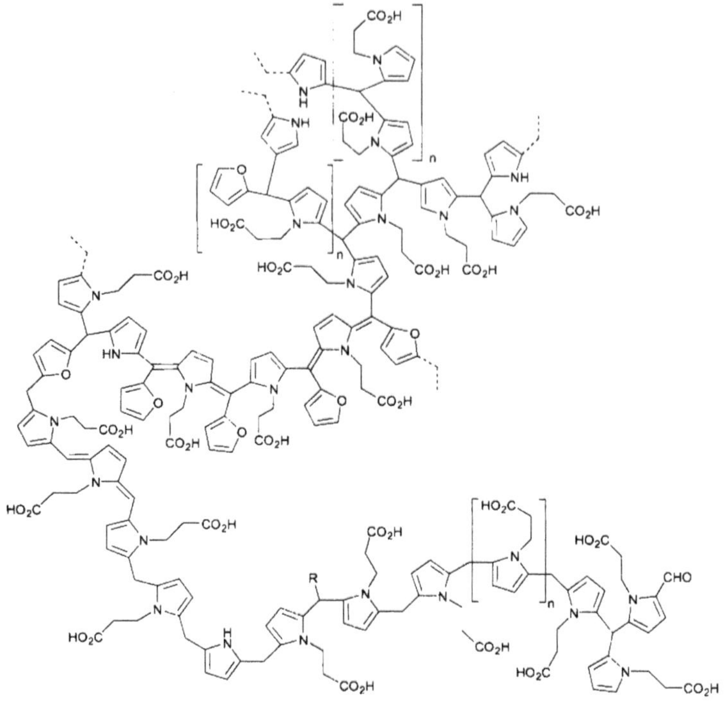

Figure 9.2   *Proposed structure of melanoidin-like polymer*[22]

Hofmann has been one of the pioneers of analysis of the Maillard reaction by NMR and has made an outstanding contribution to the field.[18,33–52] In addition to the structural analysis of Maillard reaction products derived from model systems and food extracts, his group has provided mechanistic insights into the formation of many of the identified molecules.

A powerful tool to test mechanistic ideas in complicated reactions is the use of isotopically labelled molecules in the reaction, and subsequent analysis of the products to establish the fate of the label. For example, the use of ^{13}C-labelled starting material, coupled with NMR analysis, is an extremely useful combination of methods with which to probe Maillard chemistry. A recent example of Hofmann's work is given in Case Study 9.2.

CASE STUDY 9.2 – Determining the Structure of an Intensely Orange Chromophore Resulting from the Maillard Reaction of Hexoses

In Case Study 2.2, the colour activity concept, developed by Hofmann and co-workers, was described. This method facilitates evaluation of the most intensely coloured chromophores of any Maillard reaction, without the need for knowledge of their chemical structures. This allows the challenging structural elucidation experiments to be focused on those compounds known to be responsible for producing colour.

In a recent study from the same laboratory, Frank et al.[53] used LC–MS (Chapters 5 and 6) and NMR techniques to determine the structure of one such chromophore, formed from the reaction of hexoses and primary amino acids. An intensely orange compound (Figure 9.3) was unequivocally identified as (Z)-2-[2-furylmethylidene]-5,6-di(2-furyl)-6H-pyran-3-one.

Labelling experiments using D-1-[^{13}C]glucose and D-6-[^{13}C]glucose were performed to elucidate the formation pathway of the compound, and suggested cleavage of the carbon skeleton between C5 and C6 (Figure 9.4).

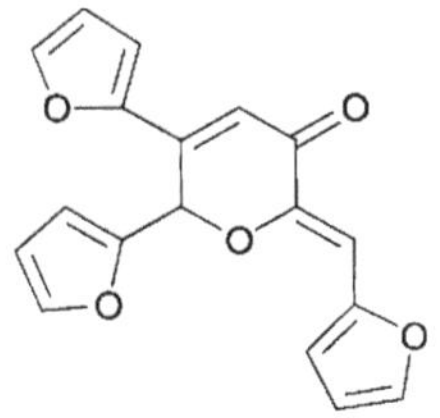

Figure 9.3   *A novel orange chromophore – (Z)-2-[2-furylmethylidene]-5,6-di(2-furyl)-6H-pyran-3-one*[53]

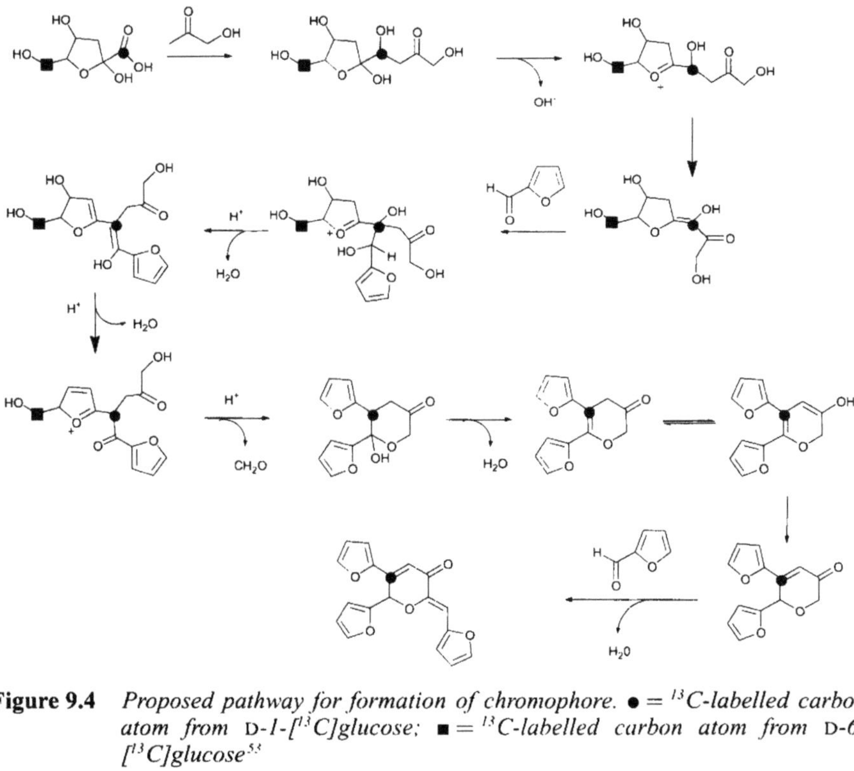

Figure 9.4   *Proposed pathway for formation of chromophore. ● = ^{13}C-labelled carbon atom from D-1-[^{13}C]glucose; ■ = ^{13}C-labelled carbon atom from D-6-[^{13}C]glucose*[53]

As the scope of Maillard research extends to the study of macromolecules, it is likely that other structure elucidation techniques more commonly employed in protein science will start to find application to determine the structure of the Maillard reaction products. For example, protein modelling has recently been used to examine the reaction of lactose with β-lactoglobulin.[54]

3   Probing an Impure Sample for Maillard Reaction Products

Immunochemical detection of specific molecules using antibodies is a technique that is finding increasing application in the medical sciences and biochemistry, where the ability to detect a specific molecule amongst a complex mixture is extremely important. So-called ELISAs – enzyme-linked immunosorbent assays – have become commonplace throughout the biological sciences, and facilitate the simple detection of any molecule against which an antibody can be raised.[55] An example of the use of such techniques is included in Case Study 9.3.

In the food sciences, immunochemical methods have yet to reach their full potential, although they are beginning to find general application in the detection of food allergens.[56-60] They have also been employed in specialist areas, such as in the detection of food moulds,[61] the identification of meat origin,[62] and for pesticide detection.[63]

The use of antibodies for the detection of Maillard products has begun to find wide application in the medical field.[17,64-71] Extensive studies have been made of immunotechniques to detect carboxymethyllysine (CML), a common product of the Maillard reaction of lysine in tissue proteins that has been used as a biomarker for various aging processes,[72 76] particularly in diabetics.[77-79] Despite this, few have realised the potential of this methodology in food analysis,[80] where there is enormous scope for detection of biomarkers in processed foods.

4   Expanding the Repertoire?

In general, reports of Maillard research in the literature tend to involve the separation and characterisation of Maillard reaction products using a limited range of techniques. Relatively few reaction products have been probed with the full repertoire of techniques available. In addition to the tenacity of researchers and the introduction of new methods to elucidate the many remaining unexplored corners of the Maillard reaction network, we predict that major advances in the field will result from combining different approaches to explore the same problem. Our final case study – an exposition of the discovery of pentosidine – is an excellent example of how this can be successfully achieved.

CASE STUDY 9.3 – In Search of Pentosidine – a Protein Crosslink Derived from Maillard Chemistry

One of the first protein-derived Maillard reaction products isolated and characterised was the crosslink pentosidine (Figure 9.5).

Pentosidine, along with CML, has served as a useful marker for Maillard damage to proteins in the medical arena,[81–97] particularly as a biomarker for diseases associated with diabetes,[98–102] renal conditions,[93,103–107] cataract formation[108–110] and of aging.[111–120]

Pentosidine is a fluorescent species[121] that is believed to form through the condensation of a lysine residue with an arginine residue and a reducing sugar. It has been studied using a number of different techniques, including instrumental methods, such as HPLC (Chapter 5),[122–125] often assisted by its intrinsic fluorescence.[81] Immunochemical techniques, such as an ELISA[94,115] have also proved immensely powerful in monitoring formation of this compound. Structural assignment, by NMR and mass spectrometry, has been assisted by synthesis of this molecule using a variety of approaches,[91,126,127] providing authentic samples for comparison to Maillard reaction product mixtures.

The exact mechanism of its formation has been the subject of considerable debate, but, aided by the substantial body of literature on this species that draws on the results of many analytical techniques, mechanisms have recently been proposed by Chellan and Nagaraj[128] and Biemel *et al.*[129] (Figure 9.6).

Compared to the extensive literature on the Maillard chemistry *in vivo*, relatively little has been reported on the existence of pentosidine in food. However, building on the results in the medical arena, Henle *et al.* have developed methods to detect the compound, and reported low levels in roasted coffee and bakery products.[130] They concluded that pentosidine does not have a major role in the polymerisation of food proteins. Iqbal *et al.* have investigated the role of pentosidine in meat tenderness in broiler hens,[131–133] and there is a report of increased pentosidine under conditions of high pressure, which could be relevant to some areas of food processing.[134]

Clearly, there is enormous scope for food researchers to adopt some of the techniques of their medical counterparts and to investigate the formation of Maillard biomarkers, such as pentosidine and CML, in a wide range of food systems. The likely benefits of such research to the food industry are enormous.

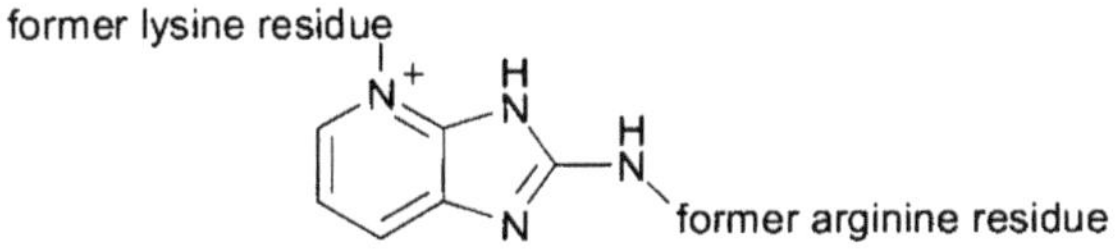

Figure 9.5 *The structure of pentosidine*

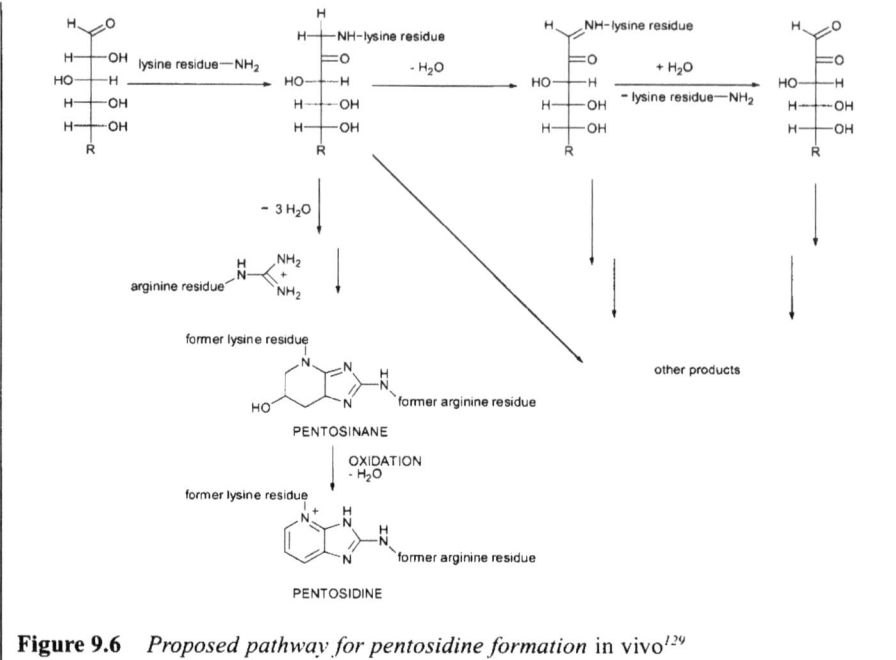

Figure 9.6 *Proposed pathway for pentosidine formation* in vivo[129]

5   In a Nutshell

We began this book by reflecting on the first reports of the Maillard reaction from Monsieur Maillard himself in 1912. If Louis-Camille were alive today, he would surely be impressed at the sheer volume of research that had been carried out bearing his name. Perhaps too, he would be delighted with the enormous potential that 21st century techniques offer to shed new light on an old problem. The scientific community is, at last, poised to elucidate the true nature of this most complex and intriguing set of reactions.

6   Further Reading

G.A. Webb, P.S. Belton, A.M. Gill and I. Delgadillo, *Magnetic Resonance in Food Science*, Royal Society of Chemistry, Cambridge, 2001.

A.P. Johnstone and M.W. Turner (eds.), *Immunochemistry – a Practical Approach*, Oxford University Press, Oxford, 1997.

7   References

1. B. KojicProdic, V. Milinkovic, J. Kidric, P. Pristovsek, S. Horvat and A. Jakas, *Carbohydr. Res.*, 1995, **279**, 21.
2. G. Suarez, M.H.L.J. Koch and A.L. Oronsky, *J. Biol. Chem.*, 1993, **268**, 17716.
3. S.R. Byrn, W. Xu and A.W. Newman, *Adv. Drug Deliv. Rev.*, 2001, **48**, 115.
4. J. Hadley, N. Malik and K. Meek, *Micron*, 2001, **32**, 307.

5. A. Barrett, M. Tsoubeli, P. Maguire, N.B. Tan, K. Conca, Y. Wang, B. Porter and I. Taub, *Cereal Chem.*, 2000, **77**, 784.
6. R. Tressl, G. Wondrak, E. Kersten and D. Rewicki, *J. Agric. Food Chem.*, 1994, **42**, 2692.
7. S. Diem and M. Herderich, *J. Agric. Food Chem.*, 2001, **49**, 2486.
8. T. Shimamura, H. Ukeda and M. Sawamura, *J. Agric. Food Chem.*, 2000, **48**, 6227.
9. R.G. Bailey, J.M. Ames and J. Mann, *J. Agric. Food Chem.*, 2000, **48**, 6240.
10. M.A. Glomb and C. Pfahler, *Carbohydr. Res.*, 2000, **329**, 515.
11. L. Pellegrino, I. De Noni and S. Cattaneo, *Nahrung*, 2000, **44**, 193.
12. J.S. Elmore and D.S. Mottram, *J. Agric. Food Chem.*, 2000, **48**, 2420.
13. K. Iiljima, M. Murata, H. Takahara, S. Irie and D. Fujimoto, *Biochem. J.*, 2000, **347**, 23.
14. A. Ravagli, G. Boschin, L. Scaglioni and A. Arnoldi, *J. Agric. Food Chem.*, 1999, **47**, 4962.
15. S. de Gennaro, G.G. Birch, S.A. Parke and B. Stancher, *Food Chem.*, 2000, **68**, 179.
16. F.J. Morales and A. Arnoldi, *Food Chem.*, 1999, **67**, 185.
17. T. Oya, N. Hattori, Y. Mizuno, S. Miyata, S. Maeda, T. Osawa and K. Uchida, *J. Biol. Chem.*, 1999, **274**, 18492.
18. T. Hofmann, *Carbohydr. Res.*, 1998, **313**, 203.
19. J.M. Ames, R.G. Bailey and J. Mann, *J. Agric. Food Chem.*, 1999, **47**, 438.
20. M.F. Wang, Y. Jin, J.G. Li and C.T. Ho, *J. Agric. Food Chem.*, 1999, **47**, 48.
21. P.C. Howard, M.I. Churchwell, L.H. Couch, M.M. Marques and D.R. Doerge, *J. Agric. Food Chem.*, 1998, **46**, 3546.
22. R. Tressl, G.T. Wondrak, L.A. Garbe, R.P. Kruger and D. Rewicki, *J. Agric. Food Chem.*, 1998, **46**, 1765.
23. A. Gomez-Sanchez, R. Paredes-Leon and J. Campora, *Magn. Reson. Chem.*, 1998, **36**, 154.
24. R. Tressl, G.T. Wondrak, R.P. Kruger and D. Rewicki, *J. Agric. Food Chem.*, 1998, **46**, 104.
25. J.M. Ames, A. Arnoldi, L. Bates and M. Negroni, *J. Agric. Food Chem.*, 1997, **45**, 1256.
26. A. Arnoldi, E.A. Corain, L. Scaglioni and J.M. Ames, *J. Agric. Food Chem.*, 1997, **45**, 650.
27. A. Keyhani and V.A. Yaylayan, *J. Agric. Food Chem.*, 1997, **45**, 697.
28. G.T. Wondrak, R. Tressl and D. Rewicki, *J. Agric. Food Chem.*, 1997, **45**, 321.
29. S. Lal, W.C. Randall, A.H. Taylor, F. Kappler, M. Walker, T.R. Brown and B.S. Szwergold, *Metab.-Clin. Exp.*, 1997, **46**, 1333.
30. J. Madaj, Y. Nishikawa, V.P. Reddy, P. Rinaldi, T. Kurata and V.M. Monnier, *Carbohydr. Res.*, 2000, **329**, 477.
31. L. Pizzoferrato, G. Rotilio and M. Paci, *J. Agric. Food Chem.*, 1999, **47**, 4060.
32. L. Pizzoferrato, M. Paci and G. Rotilio, *J. Agric. Food Chem.*, 1998, **46**, 438.
33. T. Hofmann, *Helv. Chim. Acta*, 1997, **80**, 1843.
34. T. Hofmann, *J. Agric. Food Chem.*, 1998, **46**, 3918.
35. T. Hofmann and P. Schieberle, *J. Agric. Food Chem.*, 1998, **46**, 2721.
36. T. Hofmann, *J. Agric. Food Chem.*, 1998, **46**, 3896.
37. T. Hofmann, *J. Agric. Food Chem.*, 1998, **46**, 3891.
38. T. Hofmann, *J. Agric. Food Chem.*, 1998, **46**, 3902.
39. T. Hofmann, *J. Agric. Food Chem.*, 1998, **46**, 3912.
40. T. Hofmann, *J. Agric. Food Chem.*, 1998, **46**, 932.
41. T. Hofmann, *J. Agric. Food Chem.*, 1998, **46**, 941.
42. T. Hofmann, *J. Agric. Food Chem.*, 1998, **46**, 3896.
43. T. Hofmann, *Z. Lebensm. Unters. Forsch. A*, 1998, **206**, 251.
44. T. Hofmann, *J. Agric. Food Chem.*, 1998, **46**, 3896.
45. T. Hofmann, W. Bors and K. Stettmaier, *J. Agric. Food Chem.*, 1999, **47**, 379.
46. T. Hofmann, W. Bors and K. Stettmaier, *J. Agric. Food Chem.*, 1999, **47**, 391.

47. T. Hofmann and S. Heuberger, *Food Res. Technol.*, 1999, **208**, 17.
48. T. Hofmann, *Food Res. Technol.*, 1999, **209**, 113.
49. T. Hofmann, *Eur. Food Res. Technol.*, 1999, **209**, 113.
50. T. Hofmann, *J. Agric. Food Chem.*, 1999, **47**, 4763.
51. T. Hofmann, P. Munch and P. Schieberle, *J. Agric. Food Chem.*, 2000, **48**, 434.
52. T. Hofmann and P. Schieberle, *J. Agric. Food Chem.*, 2000, **48**, 4301.
53. O. Frank, S. Heuberger and T. Hofmann, *J. Agric. Food Chem.*, 2001, **49**, 1595.
54. S.J. French and W.J. Harper, *Milk Sci. Int.*, 2001, **56**, 310.
55. A.P. Johnstone and M.W. Turner (eds.), *Immunochemistry – a Practical Approach*, Oxford University Press, Oxford, 1997.
56. S.B. Lehrer, W.E. Horner and G. Reese, *Crit. Rev. Food Sci. Nutr.*, 1996, **36**, 553.
57. S.H. Sicherer, *Pediatr. Allergy Immunol.*, 1999, **10**, 226.
58. S.B. Lehrer, L.G. Wild, K.L. Bost and R.U. Sorensen, *Clin. Rev. Allergy Immunol.*, 1999, **17**, 361.
59. S. Denery-Papini, Y. Nicolas and Y. Popineau, *J. Cereal Sci.*, 1999, **30**, 121.
60. H.A. Sampson, *J. Allergy Clin. Immunol.*, 1999, **103**, 981.
61. S.Z. Li, R.R. Marquardt and D. Abramson, *J. Food Prot.*, 2000, **63**, 281.
62. U.V. Mandokhot and S.K. Kotwal, *J. Food Sci. Technol.*, 1997, **34**, 369.
63. A. Dankwardt and B. Hock, *Food Technol. Biotechnol.*, 1997, **35**, 165.
64. Y. Sato, T. Kondo and T. Ohshima, *Histopathology*, 2001, **38**, 217.
65. N. Nyhlin, Y. Ando, R. Nagai, O. Suhr, M. El Sahly, H. Terazaki, S. Yamashita, M. Ando and S. Horiuchi, *J. Internal Med.*, 2000, **247**, 485.
66. F.A. Shamsi and R.H. Nagaraj, *Curr. Eye Res.*, 1999, **19**, 276.
67. B. Farboud, A. Aotaki-Keen, T. Miyata, L.M. Hjelmeland and J.T. Handa, *Mol. Vis.*, 1999, **5**, U1.
68. N. Miyazawa, Y. Kawasaki, J. Fujii, M. Theingi, A. Hoshi, R. Hamaoka, A. Matsumoto, N. Uozumi, T. Teshima and N. Taniguchi, *Biochem. J.*, 1998, **336**, 101.
69. T. Matsuse, E. Ohga, S. Teramoto, M. Fukayama, R. Nagai, S. Horiuchi and Y. Ouchi, *J. Clin. Pathol.*, 1998, **51**, 515.
70. K. Uchida, O.T. Khor, T. Oya, T. Osawa, Y. Yasuda and T. Miyata, *FEBS Lett.*, 1997, **410**, 313.
71. Y. Al-Abed and R. Bucala, *Bioconjugate Chem.*, 2000, **11**, 39.
72. D. Vay, M. Vidali, G. Allochis, C. Cusaro, R. Rolla, E. Mottaran, G. Bellomo and E. Albano, *Diabetologia*, 2000, **43**, 1385.
73. P. Saxena, A.K. Saxena, X.L. Cui, M. Obrenovich, K. Gudipaty and V.M. Monnier, *Invest. Ophth. Visual Sci.*, 2000, **41**, 1473.
74. K. Ikeda, R. Nagai, T. Sakamoto, H. Sano, T. Araki, N. Sakata, H. Nakayama, M. Yoshida, S. Ueda and S. Horiuchi, *J. Immunol. Methods*, 1998, **215**, 95.
75. K. Matsumoto, K. Ikeda, S. Horiuchi, H. Zhao and E.C. Abraham, *Biochem. Biophys. Res. Commun.*, 1997, **241**, 352.
76. K. fIkeda, T. Higashi, H. Sano, Y. Jinnouchi, M. Yoshida, T. Araki, S. Ueda and S. Horiuchi, *Biochemistry*, 1996, **35**, 8075.
77. C. Sady, C.L. Jiang, P. Chellan, Z. Madhun, Y. Duve, M.A. Glomb and R.H. Nagaraj, *Biochim. Biophys. Acta-Protein Struct. Molec. Enzym.*, 2000, **1481**, 255.
78. A. Abiko, M. Eto, I. Makino, N. Araki and S. Horiuchi, *Metab.-Clin. Exp.*, 2000, **49**, 567.
79. P. Shibayama, N. Araki, R. Nagai and S. Horiuchi, *Diabetes*, 1999, **48**, 1842.
80. B. Huber and M. Pischetsrieder, *J. Agric. Food Chem.*, 1998, **46**, 3985.
81. R.M. Fagugli, R. Vanholder, R. De Smet, A. Selvi, F. Antolini, N. Lameire, A. Floridi and U. Buoncristiani, *Int. J. Artif. Org.*, 2001, **24**, 256.
82. M.A. Friedlander, Y.C. Wu, C.P. Randle, G.P. Baumgardner, P.B. deOreo and V.M. Monnier in *The Maillard Reaction in Foods and Medicine*, eds. J. O'Brien, H.E. Nursten, M.J.C. Crabbe and J.N. Ames, Royal Society of Chemistry, Cambridge, 1998, 339.

83. K. Horie, T. Miyata, T. Yasuda, A. Takeda, Y. Yasuda, K. Maeda, G. Sobue and K. Kurokawa, *Biochem. Biophys. Res. Comm.*, 1997, **236**, 327.

84. D.E. Hricik, Y.C. Wu, J.A. Schulak and M.A. Friedlander, *Clin. Transplant.*, 1996, **10**, 568.

85. T. Miyata, S. Taneda, R. Kawai, Y. Ueda, S. Horiuchi, M. Hara, K. Maeda and V.M. Monnier, *Proc. Nat. Acad. Sci. U SA*, 1996, **93**, 2353.

86. V.M. Monnier and D.R. Sell in *Maillard Reactions in Chemistry, Food and Health*, eds. T.P. Laboza, G.A. Reineccius, V.M. Monnier, J. O'Brien and J.W. Baynes, Royal Society of Chemistry, Cambridge, 1994, 235.

87. R. Nagaraj, D.R. Sell, M. Prabhakaram and B. Ortweth, *Proc. Natl. Acad. Sci. USA*, 1991, **88**, 10257.

88. T. Oya, T. Osawa and S. Kawakishi, *Biosci. Biotech. Biochem.*, 1997, **61**, 263.

89. H.K. Pokharna, B. Boja, V. Monnier and R.W. Moskowitz, *Glycosylation Dis.*, 1994, **1**, 185.

90. R.A. Smulders, C.D.A. Stehouwer, C.G. Schalkwijk, A.J.M. Donker, W.M. Vanhinsbergh and J.M. Tekoppele, *Thromb. Haem.*, 1998, **80**, 52.

91. H. Sugiyama, F. Yokokawa, T. Shioiri, N. Katagiri, O. Oda and H. Ogawa, *Tetrahedron Lett.*, 1999, **40**, 2569.

92. S. Sugiyama, T. Miyata, Y. Ueda, H. Tanaka, K. Maeda, S. Kawashima, C. Van Ypersele De Strihou and K. Kurokawa, *J. Am. Soc. Nephrol.*, 1998, **9**, 1681.

93. M. Takahashi, H. Hoshino, K. Kushida, H. Murata, S. Baba and T. Inoue, *Nephron*, 1998, **80**, 444.

94. S. Taneda and V.M. Monnier, *Clin. Chem.*, 1994, **40**, 1766.

95. Y. Ueda, T. Miyata, T. Hashimoto, H. Yamada, Y. Izuhara, H. Sakai and K. Kurokawa, *Biochem. Biophys. Res. Commun.*, 1998, **245**, 785.

96. K. Yoshihara, K. Nakamura, M. Kanai, Y. Nagayama, S. Takahashi, N. Saito and M. Nagata, *Biol. Pharm. Bull.*, 1998, **21**, 1005.

97. T. Yoshimura, *Teikyo Igaku Zasshi*, 2000, **23**, 89.

98. M. Daimon, Y. Ono, T. Saito, H. Yamaguchi, A. Hirata, H. Ohnuma, M. Igarashi, H. Eguchi, H. Manaka and T. Kato, *Diabet. Care*, 1999, **22**, 877.

99. R.H. Nagaraj, T.S. Kern, D.R. Sell, J. Fogarty and R.L. Engerman, *Diabetes*, 1996, **45**, 587.

100. R.H. Nagaraj, M. Prabhakaram, B.J. Ortwerth and V.M. Monnier, *Diabetes*, 1994, **43**, 580.

101. P. Odetti, J. Fogarty, D.R. Sell and V.M. Monnier, *Diabetes*, 1992, **41**, 153.

102. D.R. Sell and V.M. Monnier, *J. Geront. A*, 1997, **52**, 284.

103. Z. Cai, T. Shinzato, Y. Matsumoto, M. Miwa, H. Otani, S. Nakai, J. Usami, H. Oka, I. Takai and K. Maeda, *Nephrology*, 1998, **4**, 407.

104. M.A. Friedlander, V. Witkosarsat, A.T. Nguyen, Y.C. Wu, M. Labrunie, C. Verger, P. Jungers and B. Descampslatscha, *Clin. Nephrol.*, 1996, **45**, 379.

105. T. Miyata, N. Ishiguro, Y. Yasuda, T. Ito, M. Nangaku, H. Iwata and K. Kurokawa, *Biochem. Biophys. Res. Comm.*, 1998, **244**, 45.

106. M.F. Weiss, R.A. Rodby, A.C. Justice and D.E. Hricik, *Kidney Int.*, 1998, **54**, 193.

107. S. Sakata, M. Takahashi, K. Kushida, M. Oikawa, H. Hoshino, M. Denda and T. Inoue, *Nephron*, 1998, **78**, 260.

108. J.T. Handa, K.M. Reiser, H. Matsunaga and L.M. Hjelmeland, *Exp. Eye Res.*, 1998, **66**, 411.

109. J.T. Handa, N. Verzijl, H. Matsunaga, A. Aotaki-Keen, G.A. Lutty, J.M.T. Koppele, T. Miyata and L.M. Hjelmeland, *Invest. Ophth. Vis. Sci.*, 1999, **40**, 775.

110. H. Hashimoto, K. Arai, S. Yoshida, M. Chikuda and Y. Obara, *Jap. J. Ophth.*, 1997, **41**, 274.

111. W.T. Cefalu, A.D. Bellfarrow, Z.Q. Wang, W.E. Sonntag, M.X. Fu, J.W. Baynes and S.R. Thorpe, *J. Geront. A*, 1995, **50**, 341.

112. J.R. Chen, M. Takahashi, M. Suzuki, K. Kushida, S. Miyamoto and T. Inoue, *J. Rheumatol.*, 1998, **25**, 2440.

113. J.R. Chen, M. Takahashi, M. Suzuki, K. Kushida, S. Miyamoto and T. Inoue, *Rheumatology*, 1999, **38**, 1275.
114. T. Kimura, J. Takamatsu, T. Miyata, T. Miyakawa and S. Horiuchi, *Pathol. Int.*, 1998, **48**, 575.
115. V.M. Monnier and D.R. Sell, *Process for Detecting Pentosidine and for Assessing the Biological Age of a Biological Sample*, PCT Int. Appl. 9707803, 1997, 127.
116. J.R. Requena, D.L. Price, S.R. Thorpe and J.W. Baynes, *Aging Methods and Protocols*, 2000, **38**, 209.
117. D.R. Sell, R.H. Nagaraj, S.K. Grandhee, P. Odetti, A. Lapolla, J. Fogarty and V.M. Monnier, *Diabetes/Metab. Rev.*, 1991, **7**, 239.
118. D.R. Sell, M. Primc, I.A. Schafer, M. Kovach, M.A. Weiss and V.M. Monnier, *Mech. Age. Dev.*, 1998, **105**, 221.
119. D.R. Sell, S. Taneda, R.H. Nagaraj, J.F. Fogarty and V.M. Monnier, *Int. Congr. Ser.*, 1995, **1100**, 280.
120. M. Takahashi, M. Oikawa and A. Nagano, *J. Geront. A*, 2000, **55**, M137.
121. D.G. Dyer, J.A. Blackledge, B.M. Katz, C.J. Hull, H.D. Adkisson, S.R. Thorpe, T.J. Lyons and J.W. Baynes, *Z. Ernaehrungswiss.*, 1991, **30**, 29.
122. Z. Deyl, I. Miksik, J. Zicha and D. Jelinkova, *Anal. Chim. Acta*, 1997, **352**, 257.
123. D.G. Dyer, J.A. Blackledge, S.R. Thorpe and J.W. Baynes, *J. Biol. Chem*, 1991, **266**, 11654.
124. M. Takahashi, H. Hoshino, K. Kushida and T. Inoue, *Anal. Biochem.*, 1995, **232**, 158.
125. M. Takahashi, H. Hoshino, K. Kushida, K. Kawana and T. Inoue, *Clin. Chem.*, 1996, **42**, 1439.
126. Y. Al-Abed and R. Bucala, *Pept. Chem., Struct. Biol., Proc. Am. Pept. Symp., 14th*, 1996, 450.
127. F. Yokokawa, H. Sugiyama, T. Shioiri, N. Katagiri, O. Oda and H. Ogawa, *Tetrahedron*, 2001, **57**, 4759.
128. P. Chellan and R.H. Nagaraj, *J. Biol. Chem.*, 2001, **276**, 3895.
129. K.M. Biemel, O. Reihl, J. Conrad and M.O. Lederer, *J. Biol. Chem.*, 2001, **276**, 23405.
130. T. Henle, U. Schwarzenbolz and H. Klostermeyer, *Food Res. Technol.*, 1997, **204**, 95.
131. M. Iqbal, L.L. Probert and H. Klandorf, *Poultry Sci.*, 1997, **76**, 1574.
132. M. Iqbal, L.L. Probert, N.H. Alhumadi and H. Klandorf, *J. Geront. A*, 1999, **54**, B171.
133. M. Iqbal, P.B. Kenney and H. Klandorf, *Poultry Sci.*, 1999, **78**, 1328.
134. U. Schwarzenbolz, H. Klostermeyer and T. Henle, *Eur. Food Res. Technol.*, 2000, **211**, 208.

Subject Index